Power Systems

For further volumes:
http://www.springer.com/series/4622

Grzegorz Benysek · Marian Pasko
Editors

Power Theories for Improved Power Quality

Springer

Grzegorz Benysek
Faculty of Electrical Engineering
Computer Science and
Telecommunications
Institute of Electrical Engineering
University of Zielona Góra
Podgórna street 50
65-246 Zielona Gora
Poland

Marian Pasko
Faculty of Electrical Engineering
Institute of Industrial Electrical
Engineering and Informatics
Silesian University of Technology
ul. Akademicka 10
44-100 Gliwice
Poland

ISSN 1612-1287 e-ISSN 1860-4676
ISBN 978-1-4471-2785-7 e-ISBN 978-1-4471-2786-4
DOI 10.1007/978-1-4471-2786-4
Springer London Heidelberg New York Dordrecht

British Library Cataloguing in Publication Data
A catalogue record for this book is available from the British Library

Library of Congress Control Number: 2012931949

Printed on acid-free paper

Springer is part of Springer Science+Business Media (www.springer.com)

Preface

Power quality is a term that describes a set of parameters of electric power and the load's ability to function properly with that electric power. Poor electric power quality can cause: overloading the network, overloading the neutral wire, dangerous resonance phenomena or even damage to the load. Generally it can lead to large economic costs particularly in countries with dynamic development of new technologies. It is estimated that problems related to power quality costs the European industry hundreds of billions of euros annually. By contrast, financing on the prevention of these problems are fragments of a percent of these costs. Therefore, research on methods of analysis and improvement of power quality are widely performed throughout the world. This book presents the issues related to methods of improving the power quality—in particular, using the active compensators. Considering the above the book can be a valuable source of information for both engineers and students in technical universities.

Chapter 1 introduces the fundamental problems of electrical power systems and elucidates the subject matter of this thesis.

Chapter 2 provides a comparison of several so-called power theories. These theories represent different approaches using both frequency and time domain. The basic criterion for the choice of the discussed theories will be historical development of knowledge in this field and the usefulness of power theory in solving practical problems: reactive power compensation, balancing the supply network load and mitigation of voltage and current distortion. Particular attention will be given to the theories defining the current components in the time domain as the basis for the present-day active compensation and filtering systems.

Chapters 3 and 4 present properties of the power electronics arrangements suitable for solving the selected power quality problems. The principle of operation and the basic properties of the series, parallel and series—parallel active compensating devices are presented.

Chapters 5 and 6 describe the operation principle of active compensators' control algorithms. Several examples of control algorithms, using the power theories described in Chap. 2 have been developed. Theoretical considerations have been illustrated by simulation and measurement results in the laboratory.

Gliwice, Poland, October 2011 Marian Pasko
Zielona Gora, Poland, October 2011 Grzegorz Benysek

Contents

Contributors

Grzegorz Benysek University of Zielona Gora, 50 Podgorna Street, 65-246 Zielona Gora, Poland, e-mail: G.Benysek@iee.zu.zgora.pl

Marcin Jarnut University of Zielona Gora, 50 Podgorna Street, 65-246 Zielona Gora, Poland, e-mail: M.Jarnut@iee.uz.zgora.pl

Jacek Kaniewski University of Zielona Gora, 50 Podgorna Street, 65-246 Zielona Gora, Poland, e-mail: J.Kaniewski@iee.uz.zgora.pl

Marcin Maciążek Silesian University of Technology, 2 Akademicka Street, 44-100 Gliwice, Poland, e-mail: marcin.maciazek@polsl.pl

Marian Pasko Silesian University of Technology, 2 Akademicka Street, 44-100 Gliwice, Poland, e-mail: marian.pasko@polsl.pl

Krzysztof Sozanski University of Zielona Gora, 50 Podgorna Street, 65-246 Zielona Gora, Poland, e-mail: K.Sozanski@iee.uz.zgora.pl

Chapter 1
Introduction

Grzegorz Benysek

Abstract The contents of this chapter encompass general problems and the most important issues of power-supply-quality improvement in AC power systems. In the context of the above, consideration is given to evaluation of bilateral interactions of loads with an electrical power distribution system and methods of their reduction. Also are discussed the basis of operation of the most important compensation-filtration devices and their applications that are joined to the system in parallel or in series.

1.1 Structure and Fundamental Problems of Electrical Power Systems

Electricity is a very useful and popular energy form which plays an increasing role in our modern industrialized society. Scarcer natural resources and the ubiquitous presence of electrical power make it desirable and continuously increase demand, causing power systems to operate close to their stability and thermal ratings. All the latter mentioned reasons together with the high penetration of distributed resources (DR) and higher than ever interest in the power quality (PQ) are the driving forces responsible for extraordinary changes taking place in the electricity supply industry worldwide.

Today's grids are primarily based on large power stations connected to transmission lines which supply power to distribution systems, thus the overall image is

G. Benysek (✉)
Institute of Electrical Engineering, University of Zielona Góra,
50 Podgórna Street, 65-246 Zielona Góra, Poland
e-mail: G.Benysek@iee.uz.zgora.pl

G. Benysek and M. Pasko (eds.), *Power Theories for Improved Power Quality*,
Power Systems, DOI: 10.1007/978-1-4471-2786-4_1,

still the same: one-way power flow from the power stations, via the transmission and distribution systems, to the final customer (end-user). Considering the above the electrical power system (EPS) can be described as a system which consists of three major components: generation, transmission and distribution. Electric power is generated at power stations predominantly by synchronous generators that are mostly driven by steam or hydro turbines. Hence, the electric power generated at any such station usually has to be transmitted over a great distance, through transmission systems to distribution systems. The distribution networks distribute the energy from the transmission grid or small/local DR to customers (end-users).

The three mentioned components—generation, transmission and distribution—have different influences, individual and sometimes common, on the level of the quality of electrical energy. There are many issues involved, such as the maintenance of power apparatus and system, the stability of the operation system, faults, distortions, loads nonlinearities etc. One must understand the potential impact of failure within one component on the performance of the whole. For example, a failure in the generation component may lead to failure in the transmission system and in a consequent loss of load in the distribution system, while a failure in the transmission component may lead to failure in the generation component and subsequent loss of customer load in distribution. A failure in the distribution system rarely leads to failure in the other two components and causes very minimal, local losses of customer load. Some of these problems are related to power transmission systems and some of them to power distribution systems, but all are fundamental from the point of view of quality of power.

From the top in the EPS hierarchy, it has to be noted that a power station which works without any failures is not a source of any difficulties in quality because the generated system voltages are almost perfectly sinusoidal. Therefore the term power quality will be treated in this thesis as a matter of two issues, related to limitations of the transmission systems [1–4] as well as to problems of the distribution systems. It is to be noted that even if PQ is mainly a distribution system problem, the power transmission system may also have an impact on the PQ issues resulting, for example, in low system damping, because of a low resistance to the reactance ratio (dynamic stability).

The PQ, at distribution level, broadly refers to maintaining a near sinusoidal power distribution bus voltage at a rated magnitude and frequency. In addition, the energy supplied to a customer must be uninterrupted. Therefore, the term power quality includes two aspects, namely Voltage Quality and Supply Reliability [5]. The Voltage Quality side includes various disturbances, such as, rapid changes, harmonics, interharmonics, flicker, imbalance and transients, whereas the reliability side involves phenomena with a longer duration, such as interruptions, voltage dips and sags, over and undervoltages and frequency deviations.

There are two different categories of causes for the deterioration in PQ, which is influenced not just by power delivery systems, but also by end-user equipment and facilities [2, 4]. The first category concerns natural causes, such as:

- faults or lightning strikes on distribution feeders;
- equipment failure.

The second category concerns load or feeder line operation:

- power electronics-based loads such as uninterrupted power supply (UPS) or Adjustable Speed Drives (ASD);
- switching on/off large loads.

This thesis builds on the assumption that interruptions and quality problems are often caused by the same phenomena, and are therefore closely related to each other; sudden and large load changes, transients, faults and loss of generation often result in the disconnection of a part of the system (reliability), while at the same time other parts experience voltage sags and short interruptions (quality problems).

An in-depth analysis of the options available for maximizing existing distribution resources, with high levels of PQ, points in the direction of power electronics [6–15]. There is general agreement that novel power electronics equipment known as Active power quality compensators (APQC) focus on the distribution system supplying the energy end-uses and is a technology created in response to reports of poor power quality of supply affecting factories, offices and homes [2, 13, 16–30]. This equipment is a potential substitute for conventional solutions, which are normally based on electro-mechanical technologies that have slow response times and high maintenance costs.

1.2 The Need for Modification

A few years back, the main concern of consumers of electricity was reliability of supply per se. It is however not only simple supply reliability that consumers want today, but they also want an ideal AC line supply, that is, a pure sine wave of fundamental frequency and, in addition, a rated peak voltage value. Unfortunately the actual AC line supply that we receive differs from this ideal. There are many ways in which the lack of quality power affects customers.

Voltage sags and dips can cause loss of production in automated processes, and can also force a computer system or data processing system to crash. To prevent such events a UPS is often used, which in turn may generate harmonics. A consumer that is connected to the same bus that supplies a large motor load may have to face a critical dip in supply voltage every time the motor load is switched on. This may be quite unacceptable to many consumers. There are also very sensitive loads, such as hospitals, air traffic control and financial institutions that require clean and uninterrupted power.

A sustained overvoltage can cause damage to household appliances. An undervoltage has the same effect as that of voltage sag. Voltage imbalance can cause temperature rises in motors. Harmonics, DC offset, can cause waveform distortions. Unwanted harmonics currents flowing across the distribution network can cause losses and heating in transformers and Electromagnetic Interference (EMI) [31–33]. Interharmonics voltages can upset the operation of fluorescent lamps and television receivers. They can also produce acoustic noise.

It can be concluded that the lack of quality power can cause loss of production and damage to equipment. It is therefore crucial that a high standard of PQ has to be maintained.

Power electronics devices can be applied to power distribution systems to increase the reliability and quality of power supplied to the customers—to increase the PQ [34–36]. The devices applied to power distribution systems for the benefit of customers (end-users) are called Active Power Quality Compensators. Through this technology the reliability and quality of the power delivered can be improved in terms of reduced interruptions and reduced voltage and current variations and distortions. The proper use of this technology will benefit all industrial, commercial and domestic customers.

APQC devices are basically used for active filtering, load balancing, power factor correction and voltage regulation. Active filtering, which predominantly is responsible for elimination of harmonic currents and voltages, can be both shunt and series. Some APQC devices are used as load compensators, in which mode they correct the imbalance and distortions in the load currents, such that compensated load draws a balanced sinusoidal current from the AC system. Some other devices are operated to provide balanced, harmonic free voltage to the customers.

1.2.1 Power Quality Issues

The term Power Quality has arisen in trying to clarify the responsibilities of utilities and customers in respect to each other, but unfortunately it is still an area of disagreement between power engineers. Many PQ-related standards are at present in existence and are under constant revision. The definition of power quality given in the Institute of Electrical and Electronic Engineers (IEEE) dictionary [37] is as follows: "Power quality is the concept of powering and grounding sensitive equipment in a matter that is suitable to the operation of that equipment."

The International Electrotechnical Commission (IEC) does not use the term Power Quality in standards, but electromagnetic compatibility and the following definition of power quality is given [38]: "The characteristics of the electricity at a given point on an electrical system, evaluated against a set of reference technical parameters—Note: These parameters might, in some cases, relate to the compatibility between electricity supplied on a network and the loads connected to that network."

A Union of the Electricity Industry (EURELECTRIC) report [39] on Power Quality in European networks states: "The quality of the electricity supply is a function of its suitability as an energy source for the electrical equipment designed to be connected to the supply network. The two primary components of supply quality are:

- continuity (freedom from interruption): the degree to which the user can rely on its availability at all times;
- voltage level: the degree to which the voltage is maintained at all times within a specified range".

[...]

"The term 'power quality' is frequently used to describe these special characteristics of the supply voltage, particularly in developed countries where discontinuity and ordinary voltage variation have largely been eliminated as matters of frequent concern. The principal phenomena concerned in power quality are:

- harmonics and other departures from the intended frequency of the alternating supply voltage;
- voltage fluctuations, especially those causing flicker;
- voltage dips and short interruptions;
- unbalanced voltages on three-phase systems;
- transient overvoltages, having some of the characteristics of high-frequency phenomena.

Power quality can be defined as the degree of any deviation from the nominal values of the abovementioned characteristics. It can be also defined as the degree to which both the utilization and delivery of electric power affects the performance of electrical equipment."

A report of the Council of European Energy Regulators (CEER) Working Group on Quality of Electricity Supply [40] states: "The main parameters of voltage quality are frequency, voltage magnitude and its variation, voltage dips, temporary or transient overvoltages and harmonic distortion. European Standard EN 50160 lists the main voltage characteristics in low and medium voltage networks, under normal operating conditions."

From all these definitions, it can be stressed that the power quality is usually considered to include two aspects of power supply, namely voltage quality and supply reliability. The voltage quality part includes different disturbances, such as rapid changes, harmonics, interharmonics, flicker, unbalance and transients; whereas the reliability part involves phenomena with a longer duration, such as interruptions, voltage dips and sags, over and undervoltages and frequency deviations. According to [3, 34] the PQ issues may be classified as in Table 1.1.

The above issues are important in describing the actual phenomena that may cause the PQ problem. Another way to categorize the different disturbances is to look at the possible causes for each kind of disturbance and to look at the consequences they might give. They are summarized in Table 1.2 [5].

1.2.1.1 Voltage Sags and Swells

A voltage sag is a short duration decrease of the root mean square (RMS) voltage, lasting from a fraction of a cycle to a few minutes in duration. These events are caused by faults on the power system or by the starting large load. Typically for transmission faults, these voltage disturbances last for fractions of a second, which represents the total fault-clearing time for transmission faults. However, these momentary events can cause a complete shutdown of plant-wide processes, which may take hours to return to normal operation.

Table 1.1 The PQ issues

Category	Spectrum	Duration	Magnitude
1.0 Transients			
1.1 Impulsive		50 ns–1 ms	<6 kV
1.2 Oscillatory		5 μs–0.3 ms	0–4 p.u.
2.0 Short duration variations			
2.1 Interruptions		10 ms–3 min	<1%
2.2 Sag		10 ms–1 min	1–90%
2.3 Swell		10 ms–1 min	110–180%
2.4 Rapid voltage changes		Not defined	>± 5%
3.0 Long duration variations		Stationary	<106% >90%
3.1 Undervoltages		>1 min	80–90%
3.2 Overvoltages		>1 min	106–120%
4.0 Voltage unbalance		Stationary	0.5–2%
5.0 Curve distortion			
5.1 DC offset	$n = 0$	Stationary	0–0.1%
5.2 Harmonics	$n = 2$–40	Stationary	0–20%
5.3 Interharmonics	0–6 kHz	Stationary	0–2%
5.4 Notches		Stationary	
5.5 Noise	Broadband	Stationary	0–1%
5.6 Signal transmission	<148 kHz	Stationary	0.09
6.0 Voltage fluctuations	<25 Hz	Intermittent	0.2–7%
7.0 Net frequency variations	50 Hz	<10 s	1%

A voltage swell occurs when a single line-to-ground fault on the system results in a temporary voltage rise on the unfaulted phases. Removing a large load or adding a large capacitor bank can also cause voltage swells, but these events tend to cause longer duration changes in the voltage magnitude and will usually be classified as long duration variations.

1.2.1.2 Voltage Interruption

A voltage interruption is the complete loss of electric voltage. Interruptions can be short duration or long duration. A disconnection of electricity causes an interruption—usually by the opening of a circuit breaker, line recloser, or fuse. For example, if a tree comes into contact with an overhead electricity line, a circuit breaker will clear the short circuit and the end-users who receive their power from the faulted line will experience an interruption. The causes of interruptions are generally the same as the causes of voltage sags and swells.

1.2.1.3 Overvoltages and Undervoltages

Long duration voltage variations that are outside the normal limits (that is, too high or too low) are most often caused by unusual conditions on the EPS. For example,

Table 1.2 Voltage disturbances

Disturbance		Origin	Consequences
Voltage sag, undervoltage	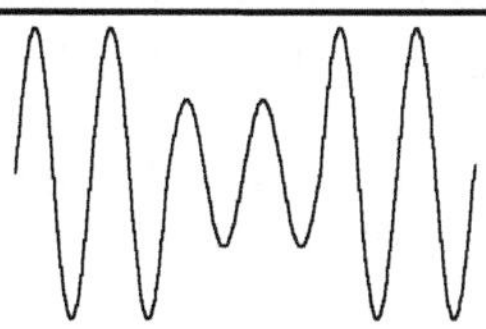	Short circuits in the network grid. Start-up of large motors	Disconnection of sensitive loads. Fail functions
Voltage swells, overvoltages	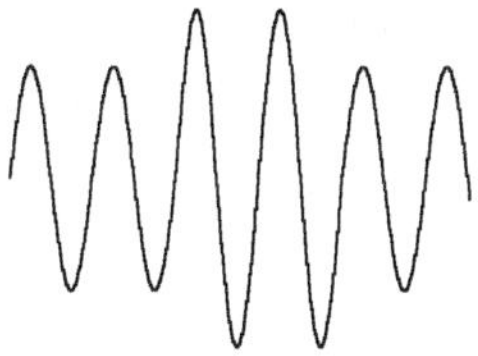	Earth fault on another phase. Shutdown of large loads. Lightning strike on network structure. Incorrect setting in substations	Disconnection of equipment may harm equipment with inadequate design margins
Harmonic distortion		Nonlinear loads. Resonance phenomena. Transformer saturation	Extended heating. Fail function of electronic equipment
Transients	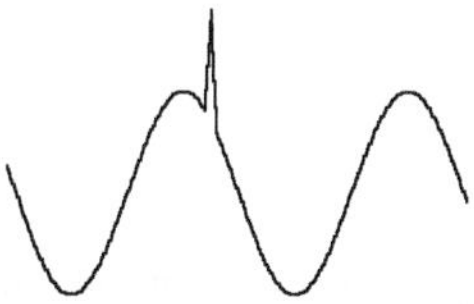	Lightning strike. Switching event	Insulation failure. Reduced lifetime of transformers, motors etc.
Voltage fluctuations, flicker	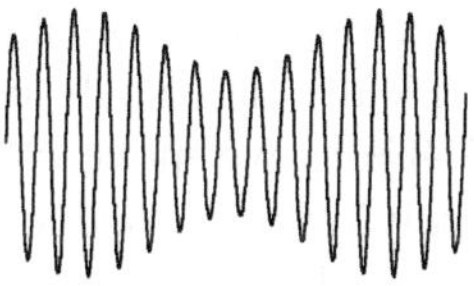	Arc furnaces. Wind turbines. Start-up of large motors	Ageing of insulation. Fail functions. Flicker
Short duration interruptions	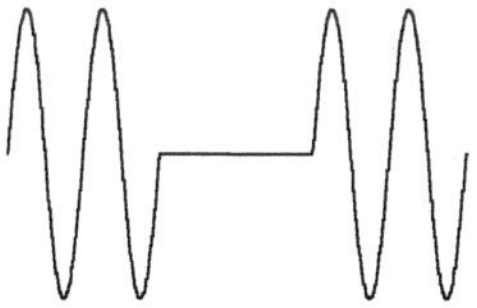	Direct short circuit. Disconnection. False tripping. Load shedding	Disconnection
Unbalanced	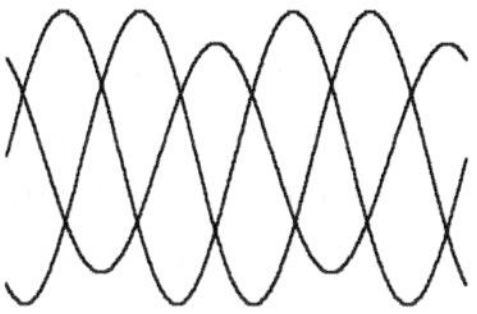	One-phase loads. Weak connections in the network	Voltage quality for overloaded phase. Overload and noise from 3-phase equipment

out-of-service lines or transformers sometimes cause undervoltage conditions. Voltage variations lasting for a longer period of time are normally corrected by adjusting the voltage with a different setting of a step voltage regulating transformer tap.

The root case of most voltage regulation problems is that there is too much impedance in the power system to properly supply the load. The load draws the current that gives a voltage drop across the system impedance. The resistive drop is in phase with the current and the reactive drop is perpendicular. Therefore, the load voltage drops low under heavy load. High voltages can come about when the source voltage is boosted to overcome the impedance drop and the load suddenly diminishes.

1.2.1.4 Voltage Flicker

A waveform may exhibit voltage flicker if its waveform amplitude is modulated at frequencies which the human eye can detect as a variation in the lamp intensity of a standard bulb. Voltage flicker is caused by an arcing condition on the EPS. Flicker problems can be corrected with the installation of filters, static VAR systems, or distribution static compensators.

1.2.1.5 Harmonic Distortion

Harmonic distortion is the presence of frequencies at integer multiples of the fundamental system frequency. Generally, it is safe to assume that the sine wave voltage generated in central power stations is pure sinusoidal. In most areas, the voltage found on transmission systems typically has much less than 1% distortion. However, the distortion may reach 5–8% as we move closer to the load. At some loads, the current waveforms will barely resemble a sine wave. Solutions to problems caused by harmonic distortion include installing active or passive filters at the load or bus, or taking advantage of transformer connections that enable cancellation of zero-sequence components.

1.2.1.6 Voltage Notching

Voltage notching is caused by the commutation of power electronic equipment. It is an effect that can raise PQ issues in any facility where solid-state rectifiers (for example, variable-speed drives) are used. The effect is caused by the switching action of the drive's input rectifier. When the drive DC link current is commutated from one rectifier thyristor to the next, an instant exists during which a line-to-line short circuit occurs at the input terminals to the rectifier.

1.3 Mitigation Methods

There are many different types of devices, which may be used to enhance the PQ, and these may be generally divided into two groups: stepwise devices and compensating type devices. Stepwise devices may regulate the voltage by use of an electronically controlled voltage tap changer, or by the use of stepwise-coupled capacitors. Such apparatus may also be used for compensation of reactive power. However, the analysis of these devices will not be performed in this thesis.

Compensating type devices usually include Voltage Source Converters (VSC) controlled by various control strategies, which, depending on the topology, may be divided into three major types: current, voltage and combined compensation.

The parallel active power filter (PAPF) may be considered the typical current compensation device, which can operate in two modes: (1) current—acts as active filter, power factor corrector, load balancer etc.; (2) voltage—regulates a bus voltage against any distortion, sag/swell, unbalance and even short duration interruptions.

Voltage-based compensation is classified as voltage harmonics filtration, voltage regulation and balancing, and removing voltage sags and dips and in general is carried out by using, e.g., series active power filter (SAPF).

Current and voltage compensation may also be combined. This combination is referred to as the unified power quality conditioner (UPQC). The conditioning functions of the UPQC are shared by the SAPF and PAPF. The SPAF performs harmonic isolation between supply and load, voltage regulation and voltage flicker/imbalance compensation, however, the PAPF performs harmonic current filtering and negative sequence balancing as well as regulation of the DC link voltage.

References

1. Hingorani N, Gyugyi L (2000) Understanding FACTS: concepts and technology of flexible ac transmission systems. IEEE, New York
2. Ghosh A, Ledwich G (2002) Power quality enhancement using custom power devices. Kluwer Academic Publishers, Boston
3. Dugan R, McGranaghan M, Beaty W (1996) Electrical power systems quality. McGraw-Hill, New York
4. Arrillaga J, Watson N, Chan S (2000) Power system quality assessment. Wiley, Chichester
5. CIGRE Working Group 14.31 (1999) Custom power–state of the art. CIGRE
6. Gyugyi L (2000) Converter-based FACTS technology: electric power transmission in the 21st century. Int Power Electron Conf 1:15–26
7. Mohan N, Undeland T, Robbins W (1995) Power electronics, converters, applications, and design, 2nd edn. Wiley, New York
8. Hingorani N (1998) Power electronics in electric utilities: role of power electronics in future power systems. Proc IEEE 76(4):481–482
9. Edris A (2000) FACTS technology development: an update. IEEE Power Eng Rev 20(3):4–9
10. Song Y, Johns A (1999) Flexible ac transmission systems (FACTS). IEE Power and Energy series 30. TJ International Ltd, Padstow
11. Hingorani N (1993) Flexible ac transmission systems. IEEE Spectrum 30(4):41–48

12. IEEE/CIGRE (1995) FACTS overview. Special issue 95-TP-108, IEEE service center, Piscataway
13. Hingorani N (1995) Introducing custom power. IEEE Spectrum 32(6):41–48
14. Akagi H (1996) New trends in active filters for power conditioning. IEEE Trans Ind App 32(6):1312–1322
15. Akagi H (1994) Trends in active power line conditioners. IEEE Trans Power Electron 9(3):263–268
16. Strzelecki R (2002) Active arrangements for energy conditioning–a new fashion or quality? (in Polish). In: Modern supplying arrangements in power systems conference, pp 1.14–9.14
17. Strzelecki R (2002) Active arrangements for energy conditioning–APC (in Polish). Przegląd Elektrotechniczny–J, 2:196–202
18. Akagi H, Fujita H (1995) A new power line conditioner for harmonic compensation in power systems. IEEE Trans Power Delivery 10(3):1570–1575
19. Akagi H (1995) New trends in active filters. In: EPE conference, pp 17–26
20. Jeon S, Cho G (1997) A series-parallel compensated uninterruptible power supply with sinusoidal input current and sinusoidal output voltage. In: IEEE-PESC conference, pp 297–303
21. Fujita H, Akagi H (1998) Unified power quality conditioner: the integration of series and shunt active filter. IEEE Trans Power Electron 13(2):315–322
22. Aredes M, Heumann K, Watanabe E (1998) An universal active power line conditioner. IEEE Trans Power Delivery 13(2):1453–1460
23. Strzelecki R, Kukluk J, Rusiński J (1999) Active power line conditioners based on symmetrical topologies. IEEE-ISIE Conf 2:825–830
24. Ghosh A, Ledwich G (2001) A unified power quality conditioner (UPQC) for simultaneous voltage and current compensation. Electr Power Syst Res 59:55–63
25. Malabika BM, Das S, Dubey G (2002) Performance study of UPQC-Q for load compensation and voltage sag mitigation. In: IEEE-IECON conference, pp 698–702
26. Meckien G, Strzelecki R (2002) Single phase active power line conditioners-without transformers. In: EPE–PEMC conference, pp 546–552
27. da Silva S (2002) A three-phase line-interactive UPS system implementation with series-parallel active power-line conditioning capabilities. IEEE Trans Ind App 38(6):1581–1590
28. Watanabe E, Aredes M (2002) Power quality considerations on shunt/series current and voltage conditioners. Conf Harmonics Qual Power 2:595–600
29. Strzelecki R (2003) New concepts of the conditioning and power flow control in the AC distribution systems. In: Modern feed equipments in electrical power systems conference, pp 65–72
30. McHattie R (1998) Dynamic voltage restorer: the customer's perspective. IEE colloquium on dynamic voltage restorer, Digest No. 98/189, Glasgow, Scotland, UK
31. Kurowski T, Benysek G, Kempski A, Smoleński R (2000) About proper cooperation of the electric drives and static converters (in Polish), vol 48. Institute of drives and electrical measurements, Wrocław University of Technology Press, Wrocław, pp 326–334
32. Strzelecki R, Kempski A, Smoleński R, Benysek G (2003) Common mode voltage cancellation in systems containing 3-phase adding transformer with PWM excitation. In: EPE Conference, pp 426–434
33. Kempski A, Strzelecki R, Smoleński R, Benysek G (2003) Suppression of conducted EMI in four-quadrant AC drive system. In: IEEE-PESC conference, pp 1121–1126
34. Thomsen P (1999) Application and control of CUPS in the distribution grid, vol 3. Institute of Energy Technology, Aalborg University, Aalborg, pp 2–11
35. Strzelecki R, Benysek G (2004) Conceptions and properties of the arrangements in distributed electrical power systems. In: MITEL conference, pp 241–248
36. Strzelecki R, Jarnut M, Benysek G (2003) Active electrical energy conditioners for individual customers. In: PES conference, vol 1. Warsaw University of Technology Press, Warsaw, pp 27–34
37. IEEE standard 100-1996 (1997) Dictionary of electrical and electronics terms, 6th edn. IEEE Standards Office New York, NY, USA

38. IEC 61000-4-30 (2003) Electromagnetic compatibility (EMC)–Part 4-30: Testing and measurement techniques–power quality measurement methods, IEC
39. EURELECTRIC (2002) Power quality in European electricity supply networks, 1st edn. Brussels Eurelectric
40. CEER working group on quality of electricity supply (2001) Quality of electricity supply: initial benchmarking on actual levels, standards and regulatory strategies, CEER

Chapter 2
Principles of Electrical Power Control

Marian Pasko and Marcin Maciążek

Abstract This chapter contains a review of the scientific works published till date in the field of power theory for systems with periodic non-sinusoidal waveforms. Nowadays, electrical energy belongs to goods indispensable in everyday life. Dynamic increase in the number of installed nonlinear loads, that are the source of higher harmonics in current and voltage waveforms, results in deterioration of electrical energy parameters. Higher harmonics make the electrical energy quality much worse. The number of power theories and papers concerning these issues give evidence about the importance of the problems of working condition optimisation in power systems.

2.1 Power Theory

Power theory is a collection of information about the properties of propagation of energy in electrical circuits. It is the result of research and experience of many generations of scientists and electrical engineers. This concept is often used in phrases such as "Fryze power theory", "instantaneous *pq* theory", etc. In this context it means a way of interpreting the phenomena occurring in the electrical system proposed by the author's ideas. The definition in this case was accompanied by the necessary formulas that permit the calculation of properties describing

---

M. Pasko (✉) · M. Maciążek
Silesian University of Technology, 2 Akademicka Street, 44-100, Gliwice, Poland
e-mail: marian.pasko@polsl.pl

M. Maciążek
e-mail: marcin.maciazek@polsl.pl

G. Benysek and M. Pasko (eds.), *Power Theories for Improved Power Quality*,
Power Systems, DOI: 10.1007/978-1-4471-2786-4_2,

the electric circuit. Power theories are also used to optimise the operating point of electrical systems. They allow to minimise losses and thereby reduce operating.

Every year, dozens of articles have been published on this subject, in one way or another trying to solve the problem of power quality. Why? The solution is purely economic, electricity is a commodity. In the market the economy which wins is the one with the better quality merchandise at a price comparable to others. The second reason is additional operating costs of the power grid. These costs are caused by:

- Increased losses in resistive elements;
- Increased losses in engines;
- Capacitor failures;
- The need to increase the efficiency of power source;
- Increased current in the neutral wire;
- Resonance phenomena (caused by higher harmonics);
- Production shutdowns caused by improper operation of protection systems.

The methods and ways of describing energy and power-quality properties related to improvement of source and load effectiveness in non-sinusoidal circuits have not been standardised so far. This is proved by the fact that in the past several decades the International Electrotechnical Commission (IEC) has changed reactive power definition several times [1–4].

2.1.1 Critical Review of Classical Power Theory, Power in Sinusoidal-Type Waveforms Circuits

The following sinusoidal waveforms are used for two-terminal networks as shown in Fig. 2.1:

$$v(t) = \sqrt{2}|V|\cos(\omega t + \alpha) \tag{2.1}$$

$$i(t) = \sqrt{2}|I|\cos(\omega t + \beta) \tag{2.2}$$

where $|V|$, $|I|$—RMS values of voltage $v(t)$ and current $i(t)$, respectively.

The different powers used in discussion of power properties of this circuit are:

- instantaneous power $p(t)$

$$\begin{aligned} p(t) = v(t)i(t) = |V||I|\cos\varphi[1 + \cos(2\omega t + 2\alpha)] \\ + |V||I|\sin\varphi\sin(2\omega t + 2\alpha) = p_1(t) + p_2(t) \end{aligned} \tag{2.3}$$

It may be expressed as:

$$p(t) = P[1 + \cos(2\omega t + 2\alpha)] + Q\sin(2\omega t + 2\alpha) \tag{2.4}$$

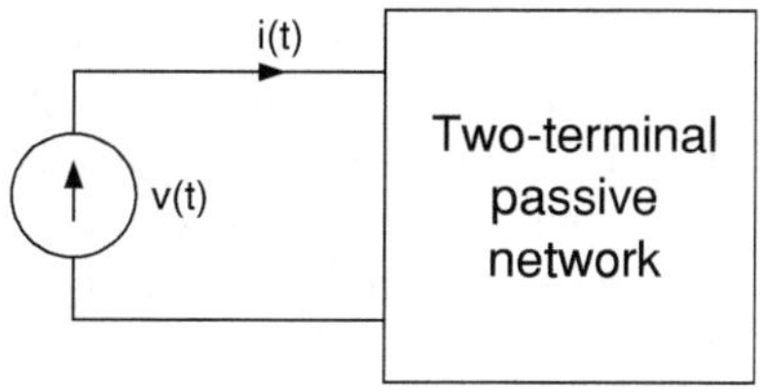

Fig. 2.1 Two-terminal network under consideration

where P—active power, Q—reactive power, φ—argument of impedance Z

$$P = \overline{p(t)} = \frac{1}{T}\int_0^T p(t)\mathrm{d}t = \frac{1}{T}\int_0^T p_1(t)\mathrm{d}t = |V||I|\cos\varphi \tag{2.5}$$

$$Q = |V||I|\sin\varphi \tag{2.6}$$

The first component of formula (2.3) describes variable non-negative component of instantaneous power with $2P$ amplitude and average value equal to load's active power P. This component represents one-directional flow of energy from the source to the load.

The second component of instantaneous power (2.3) $p_2(t)$ (alternating component) is characterised by amplitude equal to load's reactive power Q and average value equal to zero. This component characterises the bidirectional flow of energy in source-load system. It is not present if load phase angle is equal to zero. Therefore, in case of resistant load or if the load exhibits phase resonance (circuit scheme as per Fig. 2.1), two-directional oscillations in energy flow between source and load do not take place.

- Apparent power $|S|$

$$|S| = |V||I| \tag{2.7}$$

Apparent power is a purely computational quantity, it does not possess any physical meaning.

- Power factor λ

$$\lambda = \cos\varphi = \frac{P}{|S|} \tag{2.8}$$

In case of sinusoidal waveforms the power properties are described by so-called complex power:

$$S = VI^* = P + jQ \tag{2.9}$$

It must be noted that instantaneous, active, reactive and complex powers may be subjected to power balance, while apparent power may not.

All the specified powers are correctly defined, and in case of linear two-terminal network definition/interpretation is not controversial. The reactive power $Q = |V||I| \sin \varphi$ may be physically interpreted on the basis of formula (2.3) in case of one-phase linear circuits with sinusoidal waveforms. The alternating component $p_2(t)$, with amplitude equal to $Q = |V||I| \sin \varphi$ may be interpreted as the measure of backward flow of energy between circuit's reactance elements and the source. The reactive power may also be related to inductor's magnetic field or condenser's electric field. If sinusoidal current $i(t) = \sqrt{2}|I|\sin \omega t$ flows through induction coil of inductance L, magnetic field exists in the inductor and is equal to:

$$W_L(t) = \frac{1}{2}Li^2(t) = \frac{1}{2}L\left(\sqrt{2}|I|\right)^2 \sin^2 \omega t = W_{L\max} \sin^2 \omega t \tag{2.10}$$

while coil's reactive power

$$Q_L = \omega L|I|^2 = \omega W_{L\max} \tag{2.11}$$

Similarly, in case of condenser with C capacitance and supplied with sinusoidal voltage $v(t) = \sqrt{2}\,|V|\sin \omega t$, condenser's electric field energy is equal to:

$$W_C(t) = \frac{1}{2}Cv^2(t) = \frac{1}{2}C\left(\sqrt{2}|V|\right)^2 \sin^2 \omega t = W_{C\max} \sin^2 \omega t \tag{2.12}$$

and its reactive power

$$Q_C = -\omega C|V|^2 = -\omega W_{C\max} \tag{2.13}$$

Generally, in case of elements which accumulate energy, reactive power may be expressed as:

$$Q = Q_L + Q_C = \omega(W_{L\max} - W_{C\max}) \tag{2.14}$$

Compensation (reduction) of reactive power down to zero (circuit as in Fig. 2.1) minimises the RMS value of source current together with apparent power $|S|$, while active power remains unchanged; power factor goes up and attains unity. If one-phase load is non-linear, then it may be proven that reactive power does not relate in any way to energy accumulation and it may be present in purely resistance circuit [5]. Instantaneous power may also be expressed as:

$$P(t) = |V||I| \cos \varphi + |V||I| \cos(2\omega t + 2\alpha - \varphi) = P + p_P(t) \tag{2.15}$$

The first component represents active power, while the second component is alternating with amplitude equal to $|V||I|$ and corresponding to apparent power. If it is generally assumed that apparent power is a computational quantity without any physical meaning, then amplitude of alternating component defined in (2.4) may be assigned to computational quantity only. Formulas (2.4) and (2.15) show that instantaneous power may be expressed by three or two components. The number of components is influenced by mathematical approach and must not be identified with physical interpretation. We may therefore state that while instantaneous power $p(t)$ corresponds to real physical phenomena occurring in source-load

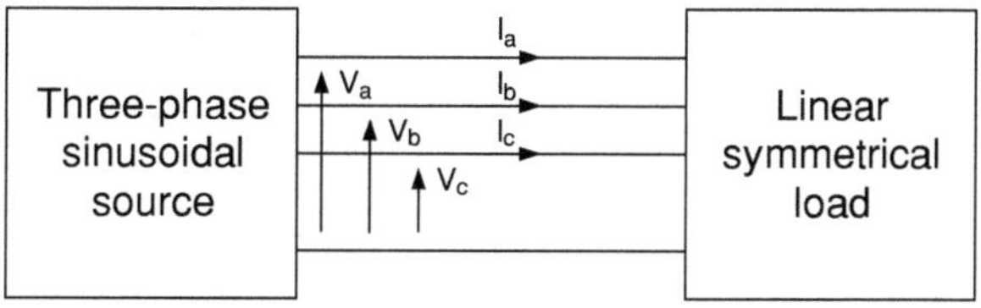

Fig. 2.2 Three-phase system under consideration

networks, the assignment of similar features to different components is, in general, not feasible. Similar interpretation of reactive power based on instantaneous power components is absolutely impossible for three-phase linear circuits in general.

For instance, if we consider symmetrical (balanced) three-phase circuit shown in Fig. 2.2.
where

$$v_a(t) = \sqrt{2}|U| \sin \omega t, \; v_b(t) = v_a\left(t - \frac{T}{3}\right), \; v_c(t) = v_a\left(t + \frac{T}{3}\right), \\ Z_a = Z_b = Z_c = |Z|e^{j\varphi} \tag{2.16}$$

instantaneous power is equal to:

$$\begin{aligned} p(t) &= v_a(t)i_a(t) + v_b(t)i_b(t) + v_c(t)i_c(t) \\ &= 2|V_a||I_a|\left[\sin \omega t \sin(\omega t - \varphi) + \sin\left(\omega t - \frac{2\pi}{3}\right)\sin\left(\omega t - \frac{2\pi}{3} - \varphi\right)\right. \\ &\quad \left. + \sin\left(\omega t + \frac{2\pi}{3}\right)\sin\left(\omega t + \frac{2\pi}{3} - \varphi\right)\right] \\ &= 3|V_a||I_a|\cos\varphi = P = \text{const} \end{aligned}$$

So, we cannot discriminate an oscillating component, which might correspond to reactive power expressed by formula:

$$Q = 3|V_a||I_a| \sin \varphi \tag{2.18}$$

On this basis alone (balanced circuit) we are able to say that there is no physical interpretation of reactive power.

To summarise: in a general case reactive power defined by formula (2.6) must be treated as some computational quantity influencing (loading) the source and decreasing its power factor. Moreover, if apparent power defined with the help of formula (2.7) for a two-terminal network is correct and not controversial, then even in case of sinusoidal three-phase networks, three different definitions of apparent power exist:

- Arithmetic apparent power [6]

$$|S_A| = |V_a||I_a| + |V_b||I_b| + |V_c||I_c| \tag{2.19}$$

- Geometric apparent power [1, 6]

$$|S_G| = \sqrt{P^2 + Q^2} \tag{2.20}$$

- Apparent power according to Buchholz [7, 8]

$$|S_B| = |\mathbf{V}||\mathbf{I}| \tag{2.21}$$

where

$|V_\alpha|, |I_\alpha|$—RMS values of source phase voltages (currents) (Fig. 2.2) $\alpha \in a, b, c$,
$P = \mathrm{Re}\{\mathbf{V}^\mathrm{T}\mathbf{I}^*\} = \mathrm{Re}\{V_a I_a^* + V_b I_b^* + V_c I_c^*\}$—active power,
$Q = \mathrm{Im}\{\mathbf{V}^\mathrm{T}\mathbf{I}^*\} = \mathrm{Im}\{V_a I_a^* + V_b I_b^* + V_c I_c^*\}$—reactive power,
$|\mathbf{V}| = \sqrt{\mathbf{V}^\mathrm{T}\mathbf{V}^*}$—RMS value of voltage vector **V**,
$|\mathbf{I}| = \sqrt{\mathbf{I}^\mathrm{T}\mathbf{I}^*}$—RMS value of current vector **I**,
$\mathbf{V}^\mathrm{T} = [V_a, V_a, V_c]$—transpose of a matrix of source phase voltage RMS phasors,
$\mathbf{I}^* = \mathrm{col}[I_a^*, I_b^*, I_c^*]$—matrix (vector) of complex conjugates of source phasor currents

Values of apparent powers defined by (2.19–2.21) differ from each other and if active power generated by the source is identical in each case, power factors λ are different. Power factor:

$$\lambda = \frac{P}{|S|} \tag{2.22}$$

may be considered to be an indicator of source usage. Only in case of symmetrical (balanced) three-phase networks the values of apparent power are identical for all definitions.

This cursory discussion demonstrates that even where linear and sinusoidal networks are concerned, there is no single uniform interpretation of different power quantities. Therefore a universally accepted "power theory" should be based on quantities with unequivocal physical interpretation in one-phase and multi-phase systems both, with sinusoidal and distorted waveforms. In our opinion, such quantities include current, voltage, their RMS values, instantaneous power, active power and—as a computational quantity—apparent power for three-phase circuits in accordance with Buchholz's formula, since it may be considered to be a natural generalisation of one-phase power concepts.

2.1.2 Budeanu Theory

In 1927 Budeanu presented his ideas of investigating power properties of circuits with non-sinusoidal waveforms. Power theory according to Budeanu [9] is at present the most widely accepted power theory of periodical and distorted waveforms; it has survived in spite of numerous opponents. Budeanu's theory owes its validity to the fact that reactive power defined thereof complies with the

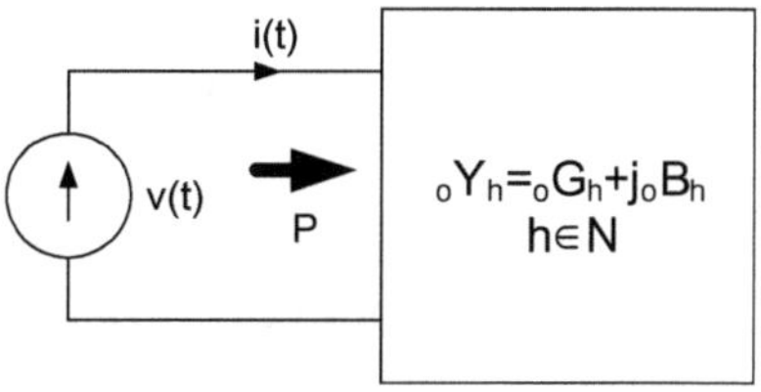

Fig. 2.3 One-phase linear circuit under consideration

power balance principle. This fact seems to point the scientists to some hidden physical interpretation of this power. Budeanu's theory is set down in every academic textbook's chapters on power phenomena in circuits with periodical and distorted waveforms. That is why we shall pay more attention to this theory here, showing both its merits and drawbacks.

In spite of numerous different approaches to power properties of distorted and periodical waveforms circuits, IEC debating in Stockholm in 1932 did not adopt any of the presented theories, since none offered generalisation features [10].

Let us return to Budeanu theory and consider the one-phase linear circuit shown in Fig. 2.3. Voltage $v(t)$ and current $i(t)$ are given in the form of Fourier series:

$$v(t) = V_0 + \sqrt{2}\mathrm{Re}\sum_{h=1}^{\infty} V_h \exp(jh\omega t) \tag{2.23}$$

$$i(t) = I_0 + \sqrt{2}\mathrm{Re}\sum_{h=1}^{\infty} I_h \exp(jh\omega t), \quad \omega = \frac{2\pi}{T} \tag{2.24}$$

where
$V_h = |V_h|\exp(j\alpha_h)$—voltage $v(t)$ RMS phasors of hth harmonic,
$I_h = |I_h|\exp(j\beta_h)$—current $i(t)$ RMS phasors of hth harmonic,
ω—pulsation of fundamental harmonic,
$\varphi_h = \beta_h - \alpha_h$—load impedance phase angle for hth harmonic.

Given this couple of waveforms $v(t)$ and $i(t)$, Budeanu has defined active power P and reactive power Q_B as superposition of active and reactive powers of all $v(t)$ and $i(t)$ harmonics:

- Active power

$$P = \frac{1}{T}\int_0^T v(t)i(t)\mathrm{d}t = V_0 I_0 + \sum_{h=1}^{\infty} |V_h||I_h|\cos\varphi_h = \sum_{h=0}^{\infty} P_h = \mathrm{Re}\sum_{h=0}^{\infty} V_h I_h^* \tag{2.25}$$

- Reactive power

$$Q_B = \sum_{h=1}^{\infty} |V_h||I_h|\sin\varphi_h = \sum_{h=1}^{\infty} Q_h = \mathrm{Im}\sum_{h=1}^{\infty} V_h I_h^* \tag{2.26}$$

- Apparent power

$$|S| = |V||I| = \sqrt{\sum_{h=0}^{\infty} |V_h|^2 \sum_{h=0}^{\infty} |I_h|^2} \tag{2.27}$$

As opposed to circuits with sinusoidal waveforms, the following inequality is true for defined powers P, Q_B and $|S|$:

$$|S|^2 \geq P^2 + Q_B^2 \tag{2.28}$$

In order to complete inequality (2.28), Budeanu has introduced a new quantity, called distortion power (he has not, however, defined the distortion concept) so that the following equality might be satisfied:

$$|S|^2 = P^2 + Q_B^2 + D_B^2 \tag{2.29}$$

Powers P, Q_B, D_B, $|S|$ may be shown graphically as so-called power rectangular prism (Fig. 2.4), and λ quantity:

$$\lambda = \cos\vartheta = \frac{P}{|S|} \tag{2.30}$$

is called a power factor.

Power theory elaborated by Budeanu has enjoyed a lot of support from the very beginning, but it has also been rejected by numerous opponents.

Budeanu's conception was severely criticised by Usatin [10] at IMEKO International Conference in Budapest in 1961. Usatin pointed to lack of physical interpretation of distortion power and unauthorised summing up of amplitudes of oscillating components of different harmonics. Moreover, he criticised the practical significance of this theory drawing attention to the fact that for 34 years no measurement device able to measure Q_B or D_B power has been constructed. Usatin thought it advisable to add squares of different components of reactive power Q_h. He also advocated the use of the forgotten Fryze's theory [5, 11].

Czarnecki discussed the matter further in his publication of 1987 [12]. He criticised Budeanu's theory showing its uselessness on the grounds that:

- apparent power cannot be minimised with the help of this theory, so that power factor cannot be increased;
- reactive power Q_B is not a measure of energy oscillations;
- reactive power does not make it possible to calculate the capacitance, whereat power factor attains highest possible value;
- there is no direct relation between current RMS value and distortion power D_B;
- independent compensation of powers Q_B and D_B is not possible;
- it implies erroneous interpretation of energy phenomena in non-sinusoidal periodical circuits.

However, Czarnecki's arguments did not convince adherents of Budeanu theory and discussion is still under way (see [13, 14] as well as the latest IEEE recommendations [1]). One of the widely used arguments in favour of applicability of

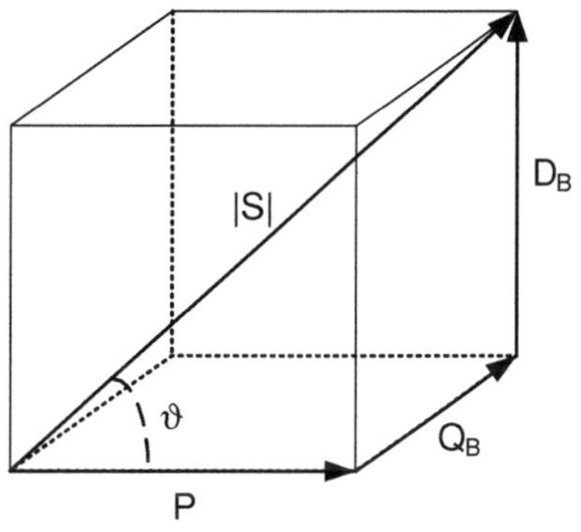

Fig. 2.4 Power rectangular prism according to Budeanu

reactive power Q_B is the fact that it is subject to energy balance (power is conserved) and—at present—it is relatively simple to design Q_B measurement devices. However, this last argument should not be considered to be substantial, if the present data processing development is taken into account.

2.1.3 Fryze Theory

In 1931 Fryze proposed a novel definition of reactive power of non-sinusoidal and periodical waveforms [5, 11].

The ruling concept was:

- first of all, for any periodical current and voltage waveform measurement of $|V|$, $|I|$, active power and power factor should be made simple; power factor was defined as:

$$\lambda = \frac{P}{|S|} = \frac{\frac{1}{T}\int_0^T v(t)i(t)\mathrm{d}t}{\sqrt{\frac{1}{T}\int_0^T v^2(t)\mathrm{d}t}\sqrt{\frac{1}{T}\int_0^T i^2(t)\mathrm{d}t}} \tag{2.31}$$

- next, to generalise the description of energy properties true for sinusoidal waveforms in such a way that they should also hold for any periodical waveforms.

We know that sinusoidal current may be decomposed into the sum of two reciprocally orthogonal components, i.e.:

$$i(t) = i_a(t) + i_b(t) \tag{2.32}$$

where

$i_a(t)$—current's active component,

$i_b(t)$—current's reactive component.

The following relation is also true for current components (it proves their orthogonality):

$$\int_0^T i_a(t)i_b(t)\mathrm{d}t = 0 \tag{2.33}$$

and:

$$\frac{1}{T}\int_0^T v(t)i(t)\mathrm{d}t = \frac{1}{T}\int_0^T v(t)i_a(t)\mathrm{d}t = P \tag{2.34}$$

Active power P may also be calculated from the expression:

$$P = |V||I_a| \tag{2.35}$$

and reactive power Q_F:

$$Q_F = |V||I_b| \tag{2.36}$$

Identical algorithm has been applied to periodical and non-sinusoidal waveforms. The following steps in reasoning can be distinguished here:

1. Axiomatic determination of active source current in accordance with equation

$$i_a(t) = {}_eGv(t) \tag{2.37}$$

where ${}_eG$—overall (equivalent) source conductance defined by the following relation:

$${}_eG = \frac{P}{\|v\|_{L^2}^2} = \frac{\frac{1}{T}\int_0^T v(t)i(t)\mathrm{d}t}{\frac{1}{T}\int_0^T v^2(t)\mathrm{d}t} = \frac{\frac{1}{T}\int_0^T v(t)i_a(t)\mathrm{d}t}{\frac{1}{T}\int_0^T v^2(t)\mathrm{d}t} \tag{2.38}$$

The active current defined here is characterised by minimum RMS value, while it ensures required power flow into the load.

2. Representation of source current as superposition of active and reactive current:

$$i(t) = i_a(t) + i_b(t) \tag{2.39}$$

but currents' orthogonality must be maintained:

$$\int_0^T i_a(t) i_b(t) \mathrm{d}t = 0 \tag{2.40}$$

3. Power definition; in case of Fryze's theory power is secondary to current decomposition:

$$\|i\|_{L^2}^2 = \|i_a\|_{L^2}^2 + \|i_b\|_{L^2}^2 \tag{2.41}$$

If we multiply both sides of Eq. 2.41 by $\|v\|^2$, the following power equation is obtained:

$$|S|^2 = P^2 + Q_F^2 \tag{2.42}$$

This reactive power $Q_F = \|v\| \|i_b\|$ has been called Fryze's reactive power. However, it must be pointed out that it is not subject to conservation.

Greater functionality of Fryze's ideas as compared to Budeanu's theory is based on the fact that decomposition into active and reactive components is carried out with primary source quantities (voltage, current), and Fourier series need not be applied here, while Budeanu's theory is based on apparent power.

Fryze was deeply opposed to the idea of elaborating power theory on the basis of Fourier series; he pointed out that taking into account Gibbs phenomenon at discontinuity points (jumps), it is not possible to minimise error produced by approximating a given function with Fourier series. Fryze's concept makes its easy to account for reactive current component, both analytically and by measurement [15, 16]. However, it does not demonstrate its physical sense, apart from the fact of excessive loading of the source. It does not provide any information about how to compensate this component with the help of two-terminal reactance networks. This current can be compensated in linear circuits by applying a controlled source with current value $i_k = -i_b$ (Fig. 2.5). This source is called active power filter.

These filters are expensive and therefore other methods of arriving at optimum system working point are used. *LC* compensators and hybrid compensators are applied [17–20].

2.1.4 Shepherd and Zakikhani Theory

Attention must be paid to the Shepherd and Zakikhani conception [21], even though its authors restricted its application to one-phase circuits. The current source has been decomposed into two components:

$$i(t) = i_R(t) + i_r(t) \tag{2.43}$$

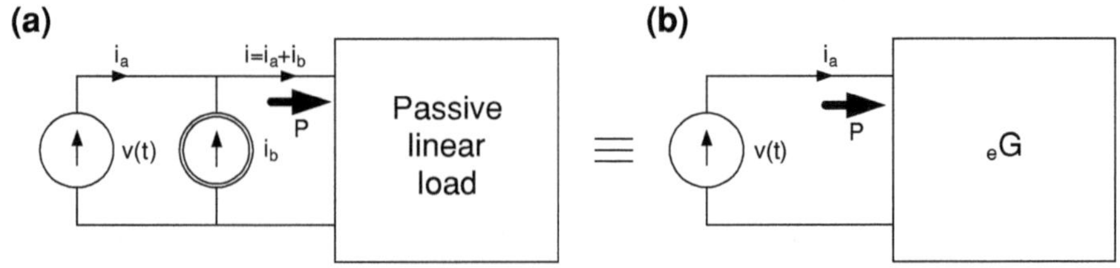

Fig. 2.5 Idea of compensation according to Fryze's theory

where

$i_R(t) = \sqrt{2}\sum_{h=1}^{\infty} |I_h| \cos\varphi_h \cos(h\omega t + \alpha_h)$—current resistance component,

$i_r(t) = \sqrt{2}\sum_{h=1}^{\infty} |I_h| \sin\varphi_h \sin(h\omega t + \alpha_h)$—current reactance component,

$\alpha_h - \arg V_h$, $\varphi_h - \angle(V_h, I_h)$,and these currents are reciprocally orthogonal.

$$\int_0^T i_R(t) i_r(t) \mathrm{d}t = 0 \tag{2.44}$$

Apparent power equation may be derived from (2.43) to (2.44):

$$|S|^2 = \|v\|^2 \|i\|^2 = \|v\|^2 \|i_R\|^2 + \|v\|^2 \|i_r\|^2 = S_R^2 + Q_r^2 \tag{2.45}$$

The active power P is not present in this equation. This was one of reasons why decomposition as per (2.43) has been severely criticised. Moreover, powers S_R and Q_r are not subject to power balance.

If we adopt decomposition into reciprocally orthogonal components (which is not always true [22]) and work out new power theories in accordance with these decompositions, then it seems that reactive power Q_r (see 2.45) as proposed by Shepherd and Zakikhani is most appropriate. Sharon [23] has modified (2.45), introducing active power as supplementary power. Apparent power equation may then be expressed as:

$|S|^2 = P^2 + Q_r^2 + S_c^2$, where $S_c^2 = S_r^2 - P^2$—supplementary power.

The authors of the above concepts did not propose any physical interpretation of S_R or S_C powers.

Interpretation of currents defined with (2.43) is easy for linear, stationary, lumped elements loads described with admittances ${}_0Y_h = {}_0G_h + j_0B_h$ (Fig. 2.3). The discussed currents may then be defined with the following relations:

$$i_R(t) = \sqrt{2}\,\mathrm{Re}\sum_{h=1}^{\infty} {}_0G_h V_h \exp(jh\omega t) \tag{2.46}$$

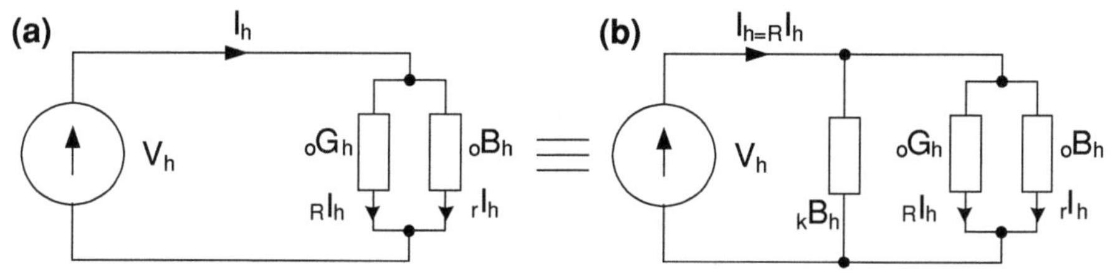

Fig. 2.6 Illustration of Shepherd's and Zakikhani's decomposition of source current

$$i_r(t) = \sqrt{2}\,\mathrm{Re}\sum_{h=1}^{\infty} j\,{}_0B_h V_h \exp(jh\omega t) \tag{2.47}$$

Current defined by (2.47) is called reactance (reactive) current and may be physically interpreted as current related to backward flow of energy (between source and load), and its measure is the reactive power Q_r. Current $i_r(t)$ may be compensated for a finite number of harmonics with the help of a two-terminal network connected to the load in parallel (Fig. 2.6b) [24]; as each considered harmonic susceptance is equal to ${}_kB_h = -{}_0B_h$. This property has been originally observed by Emanuel [24]. Basing on Shepherd and Zakikhani theory we may determine compensating capacitor's capacity, the so-called optimum capacity whereat the source factor is maximum:

$$C_{\mathrm{opt}} = \frac{\sum\limits_{h=1}^{\infty} h|V_h||I_h|\sin\varphi_h}{\omega\sum\limits_{h=1}^{\infty} h^2|V_h|^2} \tag{2.48}$$

Among the merits of this concept we can count the following:

- Definition of current $i_r(t)$, which may be compensated for a finite number of harmonics by a reactance two-terminal network;
- Determination of so-called optimum capacity value

Among its faults are:

- Active power is not present in apparent power equation;
- S_R, Q_R powers do not fit into the balance of energy;
- This theory does not cover more complex circuits than one-phase systems, even though with Sharon-added modifications active power is displayed in apparent power equation.

2.1.5 Kusters and Moore Theory

The next noteworthy theory is the one worked out by Kusters and Moore. In 1980 they published a paper presenting the main points of this concept [25]. They have decomposed source current (in case of *RL*-type load) into active current (Fryze's current), capacitative reactive current i_{qC} and residual reactive current i_{qCr}:

$$i(t) = i_a(t) + i_{qC}(t) + i_{qC_r}(t) \tag{2.49}$$

where

$$i_a(t) = \frac{P}{\|v\|^2} v(t) \quad \text{– Fryze's activecurrent,} \tag{2.50}$$

$$i_{qC}(t) = \frac{\frac{1}{T}\int\limits_0^T \frac{\mathrm{d}v}{\mathrm{d}t} i(t)\mathrm{d}t}{\left\|\frac{\mathrm{d}v}{\mathrm{d}t}\right\|^2} \frac{\mathrm{d}v}{\mathrm{d}t} \tag{2.51}$$

$$i_{qCr}(t) = i(t) - \left(i_a(t) + i_{qC}(t)\right) \tag{2.52}$$

According to their conception these currents are reciprocally orthogonal, so that RMS values fulfil the following relation:

$$\|i\|^2 = \|i_a\|^2 + \left\|i_{qC}\right\|^2 + \left\|i_{qCr}\right\|^2 \tag{2.53}$$

Power equation is expressed as:

$$|S|^2 = P^2 + Q_C^2 + Q_{Cr}^2 \tag{2.54}$$

The authors have shown that Q_C power may be fully compensated with the help of capacitor connected to the load, capacity is equal to C_{opt}:

$$C_{\mathrm{opt}} = -\frac{Q_c}{\left\|\frac{\mathrm{d}v}{\mathrm{d}t}\right\| \|v\|} \tag{2.55}$$

Similar reasoning has been proposed for *RC*-type load [25]. This concept has been quickly supported by IEC and gained a lot of popularity [4]. However, it has also been criticised and Willams [25] in particular has proved that not all Kusters-Moore statements are true.

2.1.6 Czarnecki Theory

Czarnecki has also shown in his works (1983) that Kusters and Moore theory does not satisfy all its presumed properties. Czarnecki has enriched both Fryze and Shepherd-Zakikhani concepts. His own theory is also based upon the source current decomposition into reciprocally orthogonal components. Czarnecki has exchanged Fryze's reactive current $i_b(t)$ for reactance current and scatter current:

$$i(t) = i_a(t) + i_b(t) = i_a(t) + (i_s(t) + i_r(t)) \tag{2.56}$$

Developing Shepherd-Zakikhani conception he exchanged resistance current for active current and scatter current:

$$i(t) = i_R(t) + i_r(t) = (i_a(t) + i_s(t)) + i_r(t) \tag{2.57}$$

obtaining the following decomposition:

$$i(t) = i_a(t) + i_s(t) + i_r(t) \tag{2.58}$$

If we assume that $v(t)$ may be expressed as:

$$v(t) = V_0 + \sqrt{2}\mathrm{Re}\sum_{h=1}^{\infty} V_h \exp(jh\omega t) \tag{2.59}$$

then the following relations are obtained (Fig. 2.7):

$$i_a(t) = {}_eGv(t) = {}_eGV_o + \sqrt{2}\mathrm{Re}\sum_{h=1}^{\infty} {}_eGV_h \exp(jh\omega t) \tag{2.60}$$

$$i_r(t) = \sqrt{2}\mathrm{Re}\sum_{h=1}^{\infty} j_0B_h V_h \exp(jh\omega t) \tag{2.61}$$

$$i_r(t) = \sqrt{2}\mathrm{Re}\sum_{h=1}^{\infty} j_0B_h V_h \exp(jh\omega t) \tag{2.62}$$

Since currents defined in (2.58) are reciprocally orthogonal, the following relation is also true:

$$\|i\|^2_{L^2} = \|i_a\|^2_{L^2} + \|i_r\|^2_{L^2} + \|i_s\|^2_{L^2} \tag{2.63}$$

According to Czarnecki, Eq. 2.63 explains, for the first time, why RMS current value $\|i\|$ in linear loads with periodical and non-sinusoidal voltage is greater than active current RMS value $\|i_a\|$. It is greater, when $\Lambda_h B_h \neq 0$ as well when load conductance ${}_0G_h$ varies with frequency [26]. For a finite number of harmonics current $i_r(t)$ is fully compensated with a two-terminal reactance network, but current $i_S(t)$ cannot be compensated with passive LC two-terminal network.

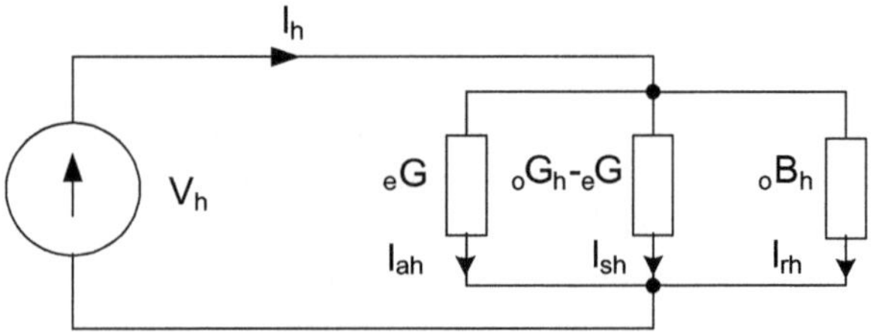

Fig. 2.7 Illustration of Czarnecki's decomposition of source

In 1991 Czarnecki stated in [27] that it is possible to compensate current with reactance-type compensator. This statement appears to be controversial. On the basis of the discussion presented in Ref. [28] we can only say that circuit shown in Fig. 2.8 may be transformed into Fig. 2.8b circuit with the help of *LC* four-terminal networks. However, optimum conditions for circuit shown in Fig. 2.8a are demonstrated in Fig. 2.8c. Figure 2.8b helps to show that source current after being compensated by the reactance-type compensator may be expressed as:

$$i_{a1} = G_a v(t) \tag{2.64}$$

and from Fig. 2.8c we can see that optimum i_a current may be expressed as:

$$i_a = {}_eGv(t) \tag{2.65}$$

In both cases source generates active power P only and source current waveform replicates voltage waveform's shape; however, minimum value of equivalent conductance G_a is dictated by ${}_0G_h$ maximum value. Since in general the following inequality is true:

$$G_a \geq ({}_0G_h)\max\ > {}_eG \tag{2.66}$$

then it is obvious that we cannot attain optimum state by this method, since $\|i_{a1}\| > \|i_a\|$.

Moreover, it has been demonstrated in Ref. [28], that $\|i_{a1}\|$ current RMS value after compensation may be greater than $\|i\|$ source current RMS value before compensation and $P_1 > P$.

Power equation must be treated as secondary product in accordance with Czarnecki conception; it may be expressed as:

$$|S|^2 = P^2 + Q_r^2 + Q_s^2 = P^2 + Q_F^2 \tag{2.67}$$

This equation will eventually lead to power rectangular prism (Fig. 2.9) different from the prism shown in Budeanu's theory; however, the sides corresponding to reactive powers Q_r and Q_S are not subject to energy balance.

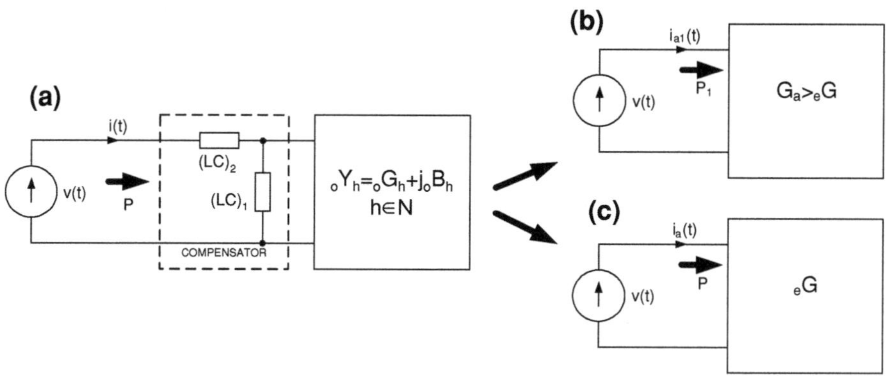

Fig. 2.8 Conception of compensating reactive and scattered currents with *LC* system [26, 28]

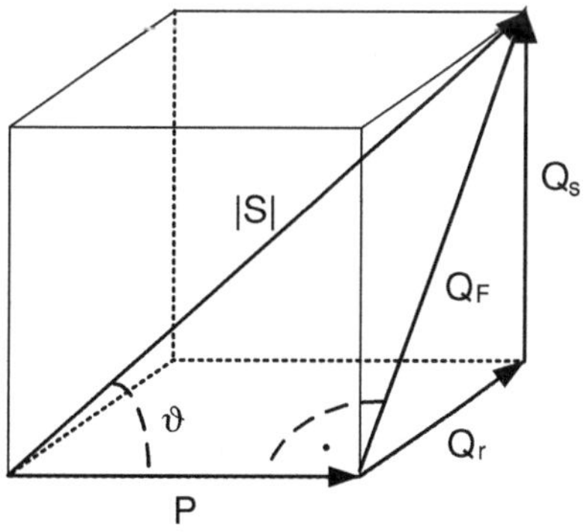

Fig. 2.9 Power rectangular prism in accordance with Czarnecki's conception

2.1.7 Optimization Theory

The idea of correlating energy and power-quality properties of a given system to solution of optimisation problem, where the input data applied uses universally accepted quantities, emerged in Institute of Electrical Circuit Theory and Engineering in 1985.

The idea may be presented as a series of claims as follows:

1. In order to characterise energy properties of non-sinusoidal circuits the following quantities are exclusively used: currents, voltages, their RMS values, instantaneous power and active power *P*
2. Optimum circuit current is defined as current calculated by solving optimisation problem with imposed side constraints
3. The optimisation quality indicator defined for a given circuit should make it possible to assess:

   (a) energy properties of waveforms—on the basis of RMS values and active power losses

(b) waveform distortion (in relation to requisite sinusoidal waveform)

4. Separate set of optimum currents defines optimum circuit condition in a given sense (by defined criteria)
5. Optimum circuit operating conditions are accomplished with the help of modifying circuits (compensators)

I. *One-phase circuits supplied from ideal periodical non-sinusoidal voltage sources*

Let us discuss the circuit shown in Fig. 2.3 with passive stationary, linear, lumped elements load. Load consumes active power P at a given voltage $e(t)$:

$$v(t) = \sqrt{2}\mathrm{Re}\sum_{h=1}^{n} |E_h| \exp(jh\omega t) \tag{2.68}$$

Source current is calculated as:

$$i(t) = \sqrt{2}\mathrm{Re}\sum_{h=1}^{n} ({}_0G_h + j{}_0\beta_h) V_h \exp(jh\omega t) \tag{2.69}$$

Optimisation problem for source operation conditions are formulated in the following way:

$$\min\|i\|_{L^2}^2 = \min\frac{1}{T}\int_0^T i(t)^2 \mathrm{d}t \tag{2.70}$$

with imposed side constraint:

$$P = (v|i)_{L^2} = \frac{1}{T}\int_0^T v(t)\, i(t) \mathrm{d}t \tag{2.71}$$

Lagrange functional was used to solve this problem. Minimisation of Lagrange functional expressed as:

$$\Phi(i, \lambda) = \| i \|_{L^2}^2 + \lambda(P - (v \,|\, i)_{L^2}) \tag{2.72}$$

results in discrimination of optimum current:

$$i^{\mathrm{opt}}(t) = i_a(t) = G_e e(t) = \sqrt{2}\mathrm{Re}\sum_{h=1}^{n} G_e V_h \exp(jh\omega t) \tag{2.73}$$

where

λ—Lagrange's multiplier

$$G_e = \frac{P}{\|v\|^2} \tag{2.74}$$

The form of optimum current coincides with Fryze's active current. Current difference:

$$i_b(t) = i(t) - i_a(t) \tag{2.75}$$

may be decomposed into reciprocally orthogonal components and may compensate different components or else current $i_b(t)$ may be compensated with the help of active filters. Complete compensation of $i_b(t)$ current helps to minimise source current RMS value, but it does not minimise current's distortion.

II. *One-phase circuits supplied from periodical non-sinusoidal voltage sources with non-zero internal impedance*

Let us discuss the circuit shown in Fig. 2.10 (for a specific harmonic), consisting of non-sinusoidal periodical voltage source with non-zero internal impedance and one-phase load.Let us assume that:

Let us assume that:

- source voltage may be expressed as:

$$e(t) = \sqrt{2}\mathrm{Re}\sum_{h=1}^{n} E_h \exp(jh\omega t) \tag{2.76}$$

- source's internal impedance belongs to stationary, linear, lumped elements class systems:

$$Z_{Zh} = R_{Zh} + jX_{Zh} \tag{2.77}$$

- load admittance belongs to stationary, linear, lumped elements class systems:

$$Y_h = G_h + jB_h \tag{2.78}$$

minimisation of Lagrange functional expressed as:

$$\Phi(i, \lambda) = \|i\|^2_{W^{2,\delta}(0,T)} + \lambda\left(P - (v|i)_{L^2(0,T)}\right) \tag{2.79}$$

results in discrimination of optimum (active) current $_a i(t)$:

$$_a i(t) = \sqrt{2}\mathrm{Re}\sum_{h=1}^{\infty} {_e G_h} E_h \exp(jh\omega t) \tag{2.80}$$

where:

$$_e G_h = \frac{\lambda_*}{2(1 + \lambda_* R_{Zh})} \tag{2.81}$$

λ_*—positive solution of equation derived from active power balance

$$P = \sum_{h=0}^{n}\left[\frac{\lambda}{2(1 + \lambda R_{Zh})} - R_h \frac{\lambda^2}{4(1 + \lambda R_{Zh})^2}\right]|E_h|^2 \tag{2.82}$$

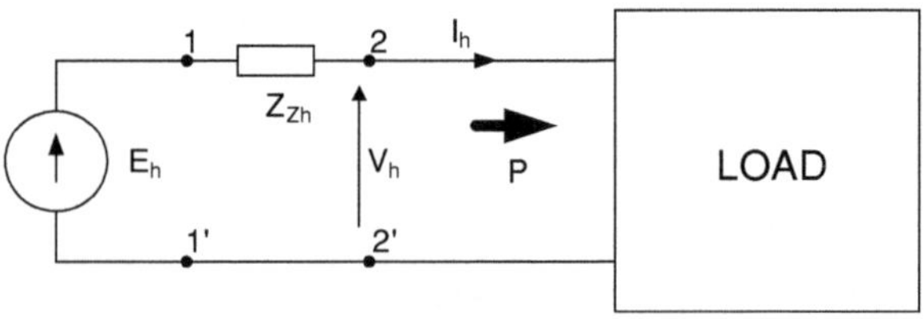

Fig. 2.10 One-phase system model for a specific (given) harmonic; source characterised by non-zero internal impedance

Taking the above into account, the remaining components may be expressed as:

- reactance component

$$_{r}i(t) = \sqrt{2}\text{Re}\sum_{h=1}^{\infty} jB_{we\,h}E_h \exp(jh\omega t) \tag{2.83}$$

- scatter component

$$_{s}i(t) = \sqrt{2}\text{Re}\sum_{h=1}^{\infty} (G_{we\,h} - {}_{e}G_h)E_h \exp(jh\omega t) \tag{2.84}$$

The author of paper [29] has proved that different current components $_{a}i(t)_{(W)}, {}_{r}\,i(t)_{(W)}, {}_{s}\,i(t)_{(W)}$ and $_{a}i(t),{}_{r}\,i(t),{}_{s}\,i(t)$ are not reciprocally orthogonal any longer, while scatter component takes part in active energy (active power) transfer. This means that elimination of one component causes changes in the remaining ones. That is why the literature of the subject [17, 30–32] proposed a different approach for sources with non-zero internal impedance. The following series of steps has to be carried out:

- discrimination of optimum current for required optimising criterion;
- calculation of compensator's current on the basis of source current (before compensation) and optimum current;
- calculation of compensator's terminal voltage on the basis of optimum current and load constants equations;
- calculation of compensator admittance on the basis of ordered pairs compensator voltage-compensator current values for specific investigated harmonics.

III. *Three-phase circuits supplied from periodical non-sinusoidal voltage sources with non-zero internal impedance*

In this section we will show how to formulate and solve exemplary optimisation problems for a selected class of three-phase circuit with a frequency approach. Three-phase system shown in Fig. 2.11 is described with the help of following data.

$$e_{\alpha}(t) = \sqrt{2}\text{Re}\sum_{h=1}^{\infty} E_{\alpha h} \exp jh\omega t \tag{2.85a}$$

$$e_b(t) = e_a\left(t - \frac{T}{3}\right), e_c(t) = e_a\left(t + \frac{T}{3}\right) \tag{2.85b}$$

internal impedance matrix of three-phase source:

$$\mathbf{Z}_h = \mathbf{R}_h + j\mathbf{X}_h\,;\ \mathbf{Z}_h = \mathbf{Z}_h^T,\ \mathrm{h} \in \mathrm{N}_0 \tag{2.86}$$

where superscript T denotes transpose of a load admittance matrix:

$$_o\mathbf{Y}_h = {_o\mathbf{G}_h} + j_o\mathbf{B}_h\,;\ _o\mathbf{Y}_h = {_o\mathbf{Y}_h^T},\ \mathrm{h} \in \mathrm{N}_0 \tag{2.87}$$

Internal impedances of both source and load are of linear, stationary, lumped elements class. Now, we will formulate the problem: **select appropriate compensators for a circuit depicted in** Fig. 2.11. Compensators admittances are defined as:

$$_k\mathbf{Y}_{\alpha 0h} = {_k\mathbf{G}_{\alpha 0h}} + j_k\mathbf{B}_{\alpha 0h},\ \alpha \in \{a, b, c\},\ \mathrm{h} \in \mathrm{N}_0 \tag{2.88}$$

Compensators should be connected between a given phase and neutral conductor (it is assumed that neutral conductor's impedance is equal to zero). This modification of the circuit should result in obtaining optimum currents as in, for instance, optimisation problem **P1**.

Problem P1. Carry out minimisation of active power losses in circuit represented by $\mathbf{R}_\mathrm{h}$:

$$\min \sum_{h=1}^{n} \mathbf{I}_h^T \mathbf{R}_h \mathbf{I}_h^* \tag{2.89}$$

Optimisation problem should be solved with three different sets of constraints related to active power P.

Variant A

$$P = \mathrm{Re}\left\{\sum_{h=1}^{n} \mathbf{E}_h^T \mathbf{I}_h^* - \sum_{h-1}^{n} \mathbf{I}_h^T \mathbf{Z}_h^T \mathbf{I}_h^*\right\} = \mathrm{const} \tag{2.90}$$

This variant is widely illustrated in references [17, 30, 32, 33], it ensures that active power remains constant in a given cross-section (see Fig. 2.11) before and after compensation. It does not ensure specific active power delivered to the load after compensation. Usually load active power after compensation is greater than before.

Variant B

$$P = \mathrm{Re}\left\{\sum_{h=1}^{n} \mathbf{V}_h^T {_o\mathbf{Y}_h^*} \mathbf{V}_h^*\right\} = \mathrm{const} \tag{2.91}$$

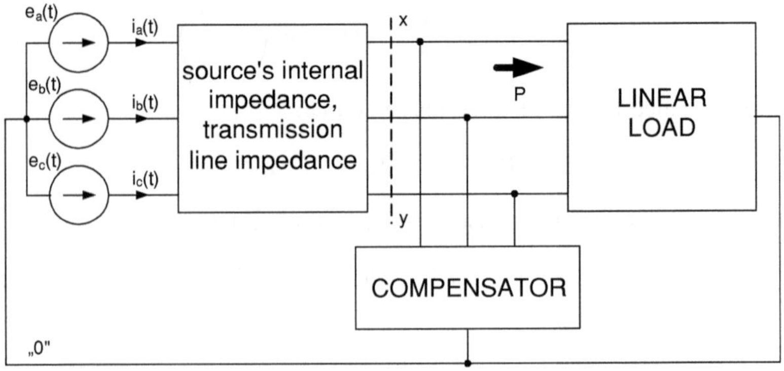

Fig. 2.11 Three-phase system under consideration

It ensures that load active power remains constant before and after compensation.

Variant C

$$\mathrm{Re}\left\{\sum_{h=1}^{n}\mathbf{E}_h^T\mathbf{I}_h^* - \sum_{h=1}^{n}\mathbf{I}_h^T\mathbf{Z}_h^T\mathbf{I}_h^*\right\} = \mathrm{Re}\left\{\sum_{h=1}^{n}\mathbf{V}_{h0}^T\mathbf{Y}_h^*\mathbf{V}_h^*\right\} \tag{2.92}$$

Compensator does not consume active power ($P_{\mathrm{komp}} = 0$). The following designations are used in (2.90–2.92):

$\mathbf{E}_h^T = [E_a, E_b, E_c]_h$—transpose of matrix of RMS phasors of source voltage's *h*th harmonic,
$\mathbf{I}_h^T = [I_a, I_b, I_c]_h$—transpose of matrix of RMS phasors of source current's *h*th harmonic,
$\mathbf{V}_h = \mathbf{E}_h - \mathbf{Z}_h\mathbf{I}_h$—matrix of load RMS phasors for *h*th harmonic,
$\mathbf{I}_h^*$—adjugate matrix of $\mathbf{I}_h$ matrix.

$$P = \sum_{\alpha=a,b,c}\frac{1}{T}\int_0^T v_\alpha(t)i_\alpha(t)\mathrm{d}t = \mathrm{Re}\left\{\sum_{h=1}^{n}\mathbf{V}_h^T I_h^*\right\},\ \alpha \in \{\mathrm{a,b,c}\} \tag{2.93}$$

Solving of different problems is carried out with Lagrange multipliers. For a given model (Fig. 2.11) the formulated problem **P1** and its variants should result in finding optimum currents of compensator currents with waveform in one phase (e.g. phase *a*).

Lagrange functional, e.g. for *variant A* of **P1** problem is expressed as:

$$\Phi(\lambda, (\mathbf{I}_h)) = \sum_{h=1}^{n}\mathbf{I}_h^T\mathbf{R}_h\mathbf{I}_h^* + \lambda\left(P - \mathrm{Re}\left\{\sum_{h=1}^{n}\mathbf{E}_h^T\mathbf{I}_h^* - \sum_{h=1}^{n}\mathbf{I}_h^T\mathbf{Z}_h^T\mathbf{I}_h^*\right\}\right) \tag{2.94}$$

Arbitrary increment $\Delta\mathbf{I}_h$ is assigned to current vector $\mathbf{I}_h$, therefore $\mathbf{I}_{1h} = \mathbf{I}_h + \Delta\mathbf{I}_h$, and:

$$\Phi_1(\lambda, (\mathbf{I}_{1h})) = \sum_{h=1}^{n} \mathbf{I}_{1h}^T \mathbf{R}_h \mathbf{I}_{1h}^* + \lambda \left(P - \mathrm{Re}\left\{ \sum_{h=1}^{n} \mathbf{E}_h^T \mathbf{I}_{1h}^* - \sum_{h=1}^{n} \mathbf{I}_{1h}^T \mathbf{Z}_h^T \mathbf{I}_{1h}^* \right\} \right) \quad (2.95)$$

Lagrange functional increment may be expressed as:

$$\Delta\Phi = \Phi_1 - \Phi = \mathrm{Re} \sum_{h=1}^{n} \left[2\mathbf{I}_h^T (\mathbf{1} + \lambda \mathbf{R}_h) - \lambda \mathbf{E}_h^T \right] \Delta\mathbf{I}_h^* + \sum_{h=1}^{n} \Delta\mathbf{I}_h^T (\mathbf{1} + \lambda \mathbf{R}_h) \Delta\mathbf{I}_h^* \quad (2.96)$$

Sufficient condition for existence of a minimum is that:

$$\bigwedge_{\Delta\mathbf{I}_h} \bigwedge_{h \in N_0} \Delta\Phi > 0 \quad (2.97)$$

Necessary condition for existence of a minimum is satisfying following vector equation:

$$\bigwedge_{h \in N_0} \left\{ 2\mathbf{I}_h^T (\mathbf{1} + \lambda \mathbf{R}_h) - \lambda \mathbf{E}_h^T \right\} = 0 \quad (2.98)$$

The sufficient condition is met for any arbitrary increment $\Delta\mathbf{I}_h$, when quadratic form denoted with (2.75) is positive, i.e.:

$$\bigwedge_{h \in N_0} \left\{ \Delta\mathbf{I}_h^T (\mathbf{1} + \lambda \mathbf{R}_h) \Delta\mathbf{I}_h^* \right\} > 0 \quad (2.99)$$

When assumptions are considered, it is seen that matrix $\mathbf{R}_h$ is symmetrical and non-singular, and therefore for every $\lambda \geq 0$ the quadratic form defined by (2.99) is positive. Coefficient $\lambda \geq 0$ is derived from power condition (2.96), where $\mathbf{I}_h(\lambda)$ current is defined as follows:

$$\mathbf{I}_h(\lambda) = \frac{\lambda}{2} (\mathbf{1} + \lambda \mathbf{R}_h)^{-1} \mathbf{E}_h \quad (2.100)$$

If we continue to act in the manner shown in [17, 29–31, 33], we obtain optimum source current for a given λ_*, and this current ensures minimisation of active power losses in the circuit:

$$\bigwedge_{h} \left(\mathbf{I}_h^{\mathrm{opt}} = \frac{\lambda_*}{2} (\mathbf{1} + \lambda_* \mathbf{R}_h)^{-1} \mathbf{E}_h \right) \quad (2.101)$$

Load voltage at optimum current (2.101) is equal to:

$$\mathbf{V}_h^{\mathrm{opt}} = \mathbf{E}_h - \mathbf{Z}_h \mathbf{I}_h^{\mathrm{opt}} = \mathbf{E}_h - \frac{\lambda_*}{2} \mathbf{Z}_h (\mathbf{1} + \lambda_* \mathbf{R}_h)^{-1} \mathbf{E}_h \quad (2.102)$$

When optimum current $\mathbf{I}_h^{\mathrm{opt}}$ and optimum voltage $\mathbf{V}_h^{\mathrm{opt}}$, $h \in N_0$ are known, the compensator's current vector ${}_k\mathbf{I}_h$ is calculated. The obtained matrix ${}_k\mathbf{I}_h$ is a column matrix, same as in the case of compensator's voltage matrix $\mathbf{V}_h^{\mathrm{opt}} = {}_k\mathbf{V}_h$. Compensator's admittance for each harmonic is calculated from the following relationship (compensator is connected between a given phase and neutral conductor):

$$ {}_kY_{\alpha h} = \frac{{}_kI_{\alpha h}}{{}_kV_{\alpha h}},\ \alpha \in \{a, b, c\} \tag{2.103}$$

Note! Theory of orthogonal distributions is true for ideal sources only.

2.2 Instantaneous Power Theories

Instantaneous power theory is based on time domain. This enables it to analyse systems in both steady-state and transient conditions. Therefore, such theories are often used in control devices to control and improve the power quality.

2.2.1 *pq* Theory

The most currently used "power theory" (in the area of improving of power quality) was presented in Japan in 1983. Instantaneous power theory proposed by Akagi and Nabae, can only be applied to the analysis of three-phase systems [34], and therefore does not have the characteristics of the general theory of power [35, 36] explain many physical phenomena), but is characterised by many advantages, of which the most important is the ability to calculate optimal current (in a given sense) with simple mathematical operations.

This theory is based on a scalar transformation of three-phase phase voltage $[v_a, v_b, v_c]^T$ and load currents $[i_{oa}, i_{ob}, i_{oc}]^T$ for the rectangular coordinate system $\boldsymbol{\alpha} - \boldsymbol{\beta} - \mathbf{0}$ (Fig. 2.12). This transformation is performed converting the instantaneous values using the formula (2.104):

$$\begin{bmatrix} F_\alpha \\ F_\beta \\ F_0 \end{bmatrix} = \sqrt{\frac{2}{3}} \begin{bmatrix} \cos \upsilon_{11} & \cos \upsilon_{12} & \cos \upsilon_{13} \\ -\sin \upsilon_{11} & -\sin \upsilon_{12} & -\sin \upsilon_{13} \\ \frac{1}{\sqrt{2}} & \frac{1}{\sqrt{2}} & \frac{1}{\sqrt{2}} \end{bmatrix} \begin{bmatrix} F_a \\ F_b \\ F_c \end{bmatrix} \tag{2.104}$$

where: υ_{1x}—angle between the axis $\mathbf{x}$ ($\mathbf{x} = \boldsymbol{a}, \boldsymbol{b}, \boldsymbol{c}$) of the natural three-phase system and the $\boldsymbol{\alpha}$ axis of the rectangular coordinate system (Fig. 2.12).

When axes $\boldsymbol{a}$ and $\boldsymbol{\alpha}$ overlap (i.e. if $\upsilon_{11} = 0$) the transformation matrix assumes the following form:

$$\begin{bmatrix} F_\alpha \\ F_\beta \\ F_0 \end{bmatrix} = \sqrt{\frac{2}{3}} \begin{bmatrix} \cos 0 & \cos \frac{4}{3}\pi & \cos \frac{2}{3}\pi \\ -\sin 0 & -\sin \frac{4}{3}\pi & -\sin \frac{2}{3}\pi \\ \frac{1}{\sqrt{2}} & \frac{1}{\sqrt{2}} & \frac{1}{\sqrt{2}} \end{bmatrix} \begin{bmatrix} F_a \\ F_b \\ F_c \end{bmatrix} \tag{2.105}$$

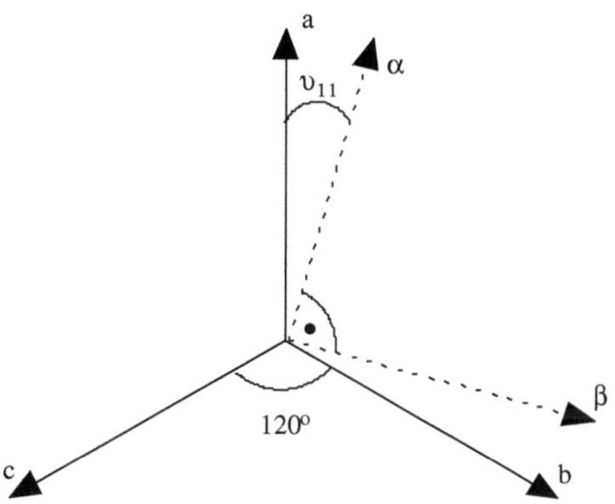

Fig. 2.12 Illustration of the transformation from $a-b-c$ to $\alpha-\beta$ coordinate system

which leads to:

$$\begin{bmatrix} F_\alpha \\ F_\beta \\ F_0 \end{bmatrix} = \sqrt{\frac{2}{3}} \begin{bmatrix} 1 & -\frac{1}{2} & -\frac{1}{2} \\ 0 & \frac{\sqrt{3}}{2} & -\frac{\sqrt{3}}{2} \\ \frac{1}{\sqrt{2}} & \frac{1}{\sqrt{2}} & \frac{1}{\sqrt{2}} \end{bmatrix} \begin{bmatrix} F_a \\ F_b \\ F_c \end{bmatrix} \tag{2.106}$$

For three-phase system of phase voltages and currents the instantaneous power expressed by the instantaneous phase currents and voltages in a ***a-b-c*** system may be determined by the formula:

$$p = v_a i_{oa} + v_b i_{ob} + v_c i_{oc} \tag{2.107}$$

After transform voltage and current phase of a three-phase system for the rectangular coordinate system $\boldsymbol{\alpha}-\boldsymbol{\beta}-\mathbf{0}$ using the formula (2.106), the instantaneous power (due to the orthogonality of transformation) preserves the form (in the new coordinate system), i.e.:

$$p = v_\alpha i_\alpha + v_\beta i_\beta + v_0 i_0 \tag{2.108}$$

In most cases, the transfer of energy in the medium voltage network is realised via three-wire line (with a symmetrical sinusoidal voltage source), then the transformation matrix can be omitted of elements v_0, i_0, and the zero component of instantaneous power.

If the p_α and p_β denote the instantaneous axis powers in $\boldsymbol{\alpha}$ and $\boldsymbol{\beta}$, then the instantaneous power can be represented as:

$$\begin{aligned} p &= p_\alpha + p_\beta = v_\alpha i_{\alpha p} + v_\alpha i_{\alpha q} + v_\beta i_{\beta p} + v_\beta i_{\beta q} \\ &= v_\alpha \frac{v_\alpha}{v_\alpha^2 + v_\beta^2} p - v_\alpha \frac{v_\beta}{v_\alpha^2 + v_\beta^2} q + v_\beta \frac{v_\beta}{v_\alpha^2 + v_\beta^2} p + v_\beta \frac{v_\alpha}{v_\alpha^2 + v_\beta^2} q \\ &= p_{\alpha p} + p_{\alpha q} + p_{\beta p} + p_{\beta q} \end{aligned} \tag{2.109}$$

where:

$i_{\alpha p}$—instantaneous active current in the α axis,
$i_{\beta p}$—instantaneous active current in the β axis,
$i_{\alpha q}$—instantaneous reactive current in the α axis,
$i_{\beta q}$—instantaneous reactive current in the β axis,
$p_{\alpha p}$—instantaneous active power in the α axis,
$p_{\alpha q}$—instantaneous reactive power in the α axis,
$p_{\beta p}$—instantaneous active power in the β axis,
$p_{\beta q}$—instantaneous reactive power in the β axis.

In this decomposition the sum of power components:

$$p_{\alpha q} + p_{\beta q} = 0 \tag{2.110}$$

These components (called instantaneous reactive powers) cancel each other and do not participate in the transfer of energy from source to load. The sum of the other two components (called instantaneous active powers):

$$p = p_{\alpha p} + p_{\beta p} \tag{2.111}$$

is identical to the standard interpretation of the instantaneous power used in the three-phase circuits, and its average value is the active power P.

Conventional passive power [9] is defined in the frequency domain and cannot in any way be compared with the instantaneous values defined in time domain (one can compare the average values for the period of the instantaneous power course with P and with Q).

The authors of the instantaneous power theory, introduced an entirely new concept—the instantaneous imaginary power (this power unit with an analogy to the *var* is marked as *vai*, or volt-ampere-imaginary). Instantaneous imaginary power is computed using the formula:

$$q = v_\alpha i_\beta - v_\beta i_\alpha \tag{2.112}$$

This component is treated as an undesirable element (it must be eliminated from the system), but it has no physical interpretation.

The main advantage of the use of rectangular coordinates $\boldsymbol{\alpha - \beta}$ is the ability to write simple equations for the phase currents. Source current of three-phase three-wire system, transformed into $\boldsymbol{\alpha - \beta}$ system can be written as:

$$\begin{bmatrix} i_\alpha \\ i_\beta \end{bmatrix} = \frac{1}{v_\alpha^2 + v_\beta^2} \begin{bmatrix} v_\alpha & -v_\beta \\ v_\beta & v_\alpha \end{bmatrix} \begin{bmatrix} \overline{p} + \tilde{p} \\ \overline{q} + \tilde{q} \end{bmatrix} \tag{2.113}$$

where $\overline{x}$—DC component of the instantaneous power, $\tilde{x}$—AC component of the instantaneous power.

For a symmetrical voltage source:

$$v_\alpha^2 + v_\beta^2 = 3|E_a|^2 = \text{const} \tag{2.114}$$

One of the tasks of optimisation may be elimination of unwanted components of the current source so that the current source was active. This task can be achieved using an active power filter. Using formula (2.115) can be calculated compensator currents (active power filter) to elimination of unnecessary parts (all or selected), leaving only the DC component of the instantaneous active power (desired), according to the formula:

$$\begin{bmatrix} i_{\alpha k} \\ i_{\beta k} \end{bmatrix} = \frac{1}{v_\alpha^2 + v_\beta^2} \begin{bmatrix} v_\alpha & -v_\beta \\ v_\beta & v_\alpha \end{bmatrix} \begin{bmatrix} p_k \\ q_k \end{bmatrix} \tag{2.115}$$

Depending on which component you want to eliminate, instead of p_k and q_k insert value from Table 2.1.

Instantaneous power theory should not be used in three-phase systems with asymmetric or distorted supply voltage *pq* theory has disadvantages as a correct power theory, as has been demonstrated in several publications [35, 36]. However, it is now often used as a useful control algorithm for active power filters.

In case where non-linear load is powered from distorted voltage source even after compensation source current will contain components related to the higher voltage harmonics. These deformations are caused by incorrect calculation of optimal currents.

2.2.2 *Extensions pq Theory*

A simple and effective method of analysis based on the *pq* theory proposed by Akagi [34] does not work in systems with unbalanced supply voltage. Since systems with small power asymmetries occur quite often we need a different approach to the analysis of such a system. In 1995, Komatsu and Kawabata [37–39] presented instantaneous power theory called the "extension *pq*". This theory uses a broader approach to ensure proper analysis of both the asymmetry and low distortion voltage.

In this theory, the instantaneous powers: active and passive are defined:

$$p = v_a i_a + v_b i_b + v_c i_c \tag{2.116}$$

$$q = v'_a i_a + v'_b i_b + v'_c i_c \tag{2.117}$$

where the transverse voltages v'_a, v'_b, v'_c are determined by the transfer phase voltages v_a, v_b, v_c separately for each phase of constant angle $\pi/2$.

In addition to the three-phase three-wire systems (such systems are the most common) can be written (according to Kirchhoff's first law):

$$i_a + i_b + i_c = 0 \tag{2.118}$$

Taking into account formula (2.118), we can simplify the formulas (2.116, 2.117) to form:

Table 2.1 Types of components to be eliminated from the instantaneous power

Component of the current source	p_k	q_k
Component associated with the instantaneous imaginary power	0	q
Component of the opposite order and higher harmonics	$\tilde{p}$	$\tilde{q}$
Component associated with the fixed component of the instantaneous imaginary power	0	$\bar{q}$
Optimum variant—full compensation components associated with the instantaneous imaginary power and higher harmonics	$\tilde{p}$	q
Component of the opposite order	$p_{2\omega}$	$q_{2\omega}$
Components associated with higher harmonics	p_h	q_h
Component associated with the variable component of the instantaneous active power	$\tilde{p}$	0

$$\begin{bmatrix} p \\ q \end{bmatrix} = \begin{bmatrix} v_a - v_c & v_b - v_c \\ v'_a - v'_c & v'_b - v'_c \end{bmatrix} \begin{bmatrix} i_a \\ i_b \end{bmatrix} \tag{2.119}$$

This formula allows you to specify the energy state of the system (calculation of instantaneous power p and q). Transforming this formula allows to calculate the value of source currents.

$$\begin{bmatrix} i_a \\ i_b \end{bmatrix} = \frac{1}{\Delta} \begin{bmatrix} v'_b - v'_c & v_c - v_b \\ v'_c - v'_a & v_a - v_c \end{bmatrix} \begin{bmatrix} p \\ q \end{bmatrix} \tag{2.120}$$
$$i_c = -i_a - i_b$$

where

$$\Delta = (v_a - v_c)(v'_b - v'_c) - (v'_a - v'_c)(v_b - v_c) \tag{2.121}$$

Analysing the form of formula (2.120) can be seen similar to the corresponding formula in the *pq* theory (2.122)

$$\begin{bmatrix} i_\alpha \\ i_\beta \end{bmatrix} = \frac{1}{\Delta'} \begin{bmatrix} v_\alpha & -v_\beta \\ v_\beta & v_\alpha \end{bmatrix} \begin{bmatrix} p \\ q \end{bmatrix} \tag{2.122}$$

where

$$\Delta' = v_\alpha^2 + v_\beta^2 \tag{2.123}$$

Currents obtained from formula (2.122) should be transformed to the ***a-b-c*** system. At the same time, you can prove that this theory is equivalent to *pq* theory in the field of symmetrical voltage power system, i.e. if:

$$v_a + v_b + v_c = 0 \tag{2.124}$$

Then the transverse voltages can be written as:

$$v'_a = \frac{v_c - e_b}{\sqrt{3}} \tag{2.125}$$

$$v_b' = \frac{v_a - v_c}{\sqrt{3}} \tag{2.126}$$

$$v_c' = \frac{v_b - v_a}{\sqrt{3}} \tag{2.127}$$

and phase current in phase a can be written as:

$$i_a = \frac{2v_b p + v_a(-p + \sqrt{3}q) - v_c(p + \sqrt{3}q)}{2(v_a^2 + v_b^2 - v_b v_c + v_c^2 - v_a(v_b + v_c))} \tag{2.128}$$

By contrast, given that in pq theory:

$$v_\alpha = \sqrt{\frac{2}{3}}\left(v_a - \frac{1}{2}v_b - \frac{1}{2}v_c\right) \tag{2.129}$$

$$v_\beta = \sqrt{\frac{2}{3}}\left(\frac{\sqrt{3}}{2}v_b - \frac{\sqrt{3}}{2}v_c\right) \tag{2.130}$$

this current phase a (in theory pq) can be written as:

$$i_a = \frac{2v_b p + v_a(-p + \sqrt{3}q) - v_c(p + \sqrt{3}q)}{2(v_a^2 + v_b^2 - v_b v_c + v_c^2 - v_a(v_b + v_c))} \tag{2.131}$$

Comparing formulas (2.128) and (2.131) can be seen that they are identical. Proceeding similarly with the currents in other phases we receive the same results. This demonstrates the broader approach in the theory of "extension pq", and the equivalence of these theories in the case of symmetric voltages source (because the formulas (2.125–2.127) are true in this case).

2.2.3 Synchronous Reference Frame Theory

This generalisation of the pq theory should be applied especially in cases of deformed voltage systems. This approach uses the transformation [40] vectors of the input signal from the natural three-phase system to a rotating $\boldsymbol{d-q}$ coordinate system as shown in Fig. 2.13.

Transformation is carried out in two stages, the first stage transforms the vectors of the three-phase system to rectangular coordinate system $\boldsymbol{\alpha} - \boldsymbol{\beta}$ similarly as in pq theory [34], i.e. according to dependence:

$$\begin{bmatrix} F_\alpha \\ F_\beta \end{bmatrix} = \sqrt{\frac{2}{3}} \begin{bmatrix} 1 & -\frac{1}{2} & -\frac{1}{2} \\ 0 & \frac{\sqrt{3}}{2} & -\frac{\sqrt{3}}{2} \end{bmatrix} \begin{bmatrix} F_a \\ F_b \\ F_c \end{bmatrix} \tag{2.132}$$

In the next step vectors are transformed into $\boldsymbol{d}-\boldsymbol{q}$ rotating system. Values in the new coordinate system are derived from the dependence:

$$\overrightarrow{F}_{dq}=\overrightarrow{F}_{\alpha\beta}\,e^{-j\theta} \tag{2.133}$$

leads to:

$$\begin{bmatrix} F_d \\ F_q \end{bmatrix}=\begin{bmatrix} \cos\theta & \sin\theta \\ -\sin\theta & \cos\theta \end{bmatrix}\begin{bmatrix} F_\alpha \\ F_\beta \end{bmatrix} \tag{2.134}$$

This transformation in the literature is also called Park transformation.

Similarly, the inverse transformation in the first place requires a transformation of vectors from the $\boldsymbol{d}-\boldsymbol{q}$ rotating to rectangular $\boldsymbol{\alpha}-\boldsymbol{\beta}$ coordinate system:

$$\begin{bmatrix} F_\alpha \\ F_\beta \end{bmatrix}=\begin{bmatrix} \cos\theta & -\sin\theta \\ \sin\theta & \cos\theta \end{bmatrix}\begin{bmatrix} F_d \\ F_q \end{bmatrix} \tag{2.135}$$

Then the transformation to the natural three-phase system:

$$\begin{bmatrix} F_a \\ F_b \\ F_c \end{bmatrix}=\sqrt{\frac{2}{3}}\begin{bmatrix} 1 & 0 \\ -\frac{1}{2} & \frac{\sqrt{3}}{2} \\ -\frac{1}{2} & -\frac{\sqrt{3}}{2} \end{bmatrix}\begin{bmatrix} F_\alpha \\ F_\beta \end{bmatrix}. \tag{2.136}$$

The advantage of this transformation is that the system does not require calculation of instantaneous active and reactive power. If the course function $\cos\theta$ coincides with the course of the basic harmonic voltage phase a, then the system dq rotates synchronously with the basic harmonic voltage course. In this reference system components that are in phase with the fundamental harmonic voltage (e.g., active current), are represented by constant values. In this case, the mean value of current component of $\boldsymbol{d}$-axis (Fig. 2.13) corresponds to the active component of source current in the rotating coordinate system [41]. Therefore, the optimal values of instantaneous currents can be calculated directly from knowledge of the components of phase currents in the new coordinate system (**d** and **q**). This follows from the fact that the only desired component is a constant (DC) component of i_d:

$$\begin{bmatrix} i_{ka} \\ i_{kb} \\ i_{kc} \end{bmatrix}=\sqrt{\frac{2}{3}}\begin{bmatrix} 1 & 0 \\ -\frac{1}{2} & \frac{\sqrt{3}}{2} \\ -\frac{1}{2} & -\frac{\sqrt{3}}{2} \end{bmatrix}\begin{bmatrix} \cos\theta & -\sin\theta \\ \sin\theta & \cos\theta \end{bmatrix}\begin{bmatrix} \tilde{i}_d \\ i_q \end{bmatrix} \tag{2.137}$$

where $\tilde{i}_d$—AC component of i_d current component.

Eliminate the calculation of instantaneous active and reactive power values, causing significant reduction in the number of required mathematical operations in each calculation cycle. This leads to improved dynamic properties of the system. Use PLL to determine θ angle ensures correct calculation of the optimal currents and compensation currents, even in the case of distorted periodic voltage. In the case of a symmetric voltage, this method is equivalent to the method proposed by Akagi [34].

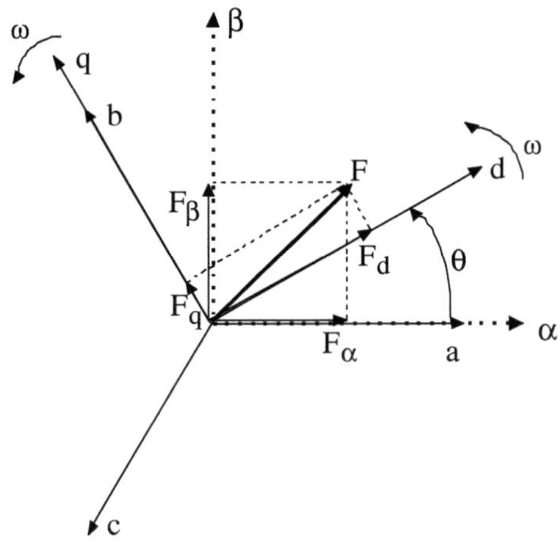

Fig. 2.13 Illustration of transformation from a three-phase $a - b - c$ system to rotating $d - q$ system

2.3 Power Theory on Basis of Orthogonal Components

The CPC theory [42] proposed by L. Czarnecki, uses a frequency domain for description of system working point. In this theory source current orthogonal decomposition has been used. These components are referred to physical interpretation for the phenomena in electrical circuits. Basic to this theory for single-phase systems have been presented already in 1984, a complete theory for the different supply conditions and different loads was published in 1994. It combines the ideas proposed by Fryze (time domain) and Shepherd and Zakikhani (frequency domain). Power theory developed by Czarnecki select from these theories this elements that allow to explain energy phenomenas in electrical circuits and establish the theoretical basis of their compensation. To describe current and voltages waveforms Fourier series are used, for example phase voltage is expressed in the form of a column vector **u**:

$$\mathbf{V} = \begin{bmatrix} v_a \\ v_b \\ v_c \end{bmatrix} = \sqrt{2}\mathrm{Re}\begin{bmatrix} \underline{V}_a \\ \underline{V}_b \\ \underline{V}_c \end{bmatrix} e^{\mathrm{j}\omega t} = \sqrt{2}\mathrm{Re}\mathbf{V}e^{\mathrm{j}\omega t} \tag{2.138}$$

Phase currents are presented, analogous to voltage, using a column vector **i**:

$$\mathbf{i} = \begin{bmatrix} i_a \\ i_b \\ i_c \end{bmatrix} = \sqrt{2}\mathrm{Re}\begin{bmatrix} \underline{I}_a \\ \underline{I}_b \\ \underline{I}_c \end{bmatrix} e^{\mathrm{j}\omega t} = \sqrt{2}\mathrm{Re}\mathbf{I}e^{\mathrm{j}\omega t} \tag{2.139}$$

In the example system shown in Fig. 2.14 (with sinusoidal voltages and currents) current source can be presented as:

$$\mathbf{i} = \sqrt{2}\mathrm{Re}\{[G_e + jB_e)\mathbf{V} + A\mathbf{V}^{\#}]e^{j\omega t} \quad (2.140)$$

where

$$G_e + jB_e = Y_{ab} + Y_{bc} + Y_{ac} \equiv Y_e \quad (2.141)$$

Y_e—equivalent admittance.

$$A \equiv -(Y_{bc} + \alpha Y_{ac} + \alpha^* Y_{ab}) \quad (2.142)$$

In this case, the author of CPC power theory proposed a source decomposition into three orthogonal components:

$$\mathrm{i} = \mathrm{i}_a + \mathrm{i}_r + \mathrm{i}_u \quad (2.143)$$

Each of them is related to another energy phenomenon, and so:

- Component $\mathbf{i}_a$ called “active current”

$$\mathbf{i}_a = \sqrt{2}\mathrm{Re}\{G_e \mathbf{V} e^{j\omega t}\} \quad (2.144)$$

is related to permanent energy conversion.

- Component $\mathbf{i}_\mathrm{r}$ called “reactive current”

$$\mathbf{i}_r = \sqrt{2}\mathrm{Re}\{B_e \mathbf{V} e^{j\omega t}\} \quad (2.145)$$

is related to phase-shift.

- Component $\mathbf{i}_\mathrm{u}$ called “unbalanced current”

$$\mathbf{i}_u = \sqrt{2}\mathrm{Re}\{A\mathbf{V}^{\#} e^{j\omega t}\} \quad (2.146)$$

is related to load imbalance.

If a system contains non-linear loads and waveforms of currents and voltages are distorted, then the supply voltage is described by the following equation:

$$\mathbf{V} = \begin{bmatrix} v_a \\ v_b \\ v_c \end{bmatrix} = \sqrt{2}\mathrm{Re}\sum_{n \in N} \begin{bmatrix} \underline{V}_a \\ \underline{V}_b \\ \underline{V}_c \end{bmatrix} e^{\mathrm{j}n\omega_1 t} = \sqrt{2}\mathrm{Re}\sum_{n \in N} \underline{\mathbf{V}}_n e^{\mathrm{j}n\omega_1 t} \quad (2.147)$$

Thus, according to the theory of current physical components, source current was in this case, distributed into four components:

$$\mathbf{i} = \mathbf{i}_a + \mathbf{i}_s + \mathbf{i}_r + \mathbf{i}_u \quad (2.148)$$

These components are mutually orthogonal, i.e.:

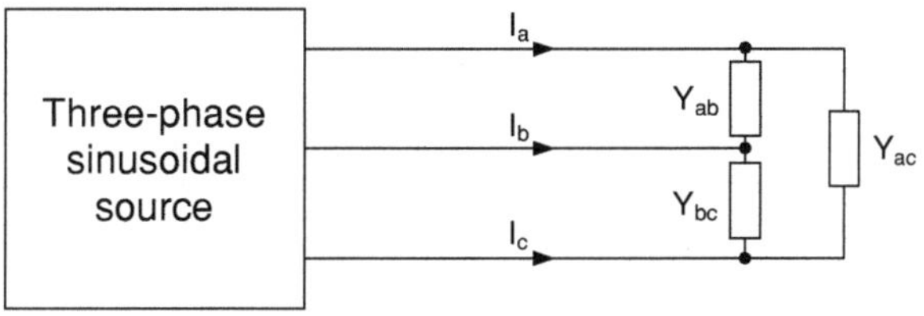

Fig. 2.14 Example system with three-phase load

$$\|\mathbf{i}\|^2 = \|\mathbf{i}_a\|^2 + \|\mathbf{i}_s\|^2 + \|\mathbf{i}_r\|^2 + \|\mathbf{i}_u\|^2 \tag{2.149}$$

And each of them is related to another energy phenomenon:

$$\mathbf{i}_a = \sqrt{2}\mathrm{Re}\left\{\sum_{n \in N} G_e \mathbf{V}_n e^{jn\omega_1 t}\right\} \tag{2.150}$$

Active current is responsible for the flow of energy from a power source to the receiver. This occurs when the receiver has non-zero active power:

$$\mathbf{i}_s = \sqrt{2}\mathrm{Re}\left\{\sum_{n \in N} (G_{en} - G_e)\mathbf{V}_n e^{jn\omega_1 t}\right\} \tag{2.151}$$

This component (scattered current) does not participate in energy flow from a source to the load. It is due to change in load conductance G_{en} with a row of harmonic n.

$$\mathbf{i}_r = \sqrt{2}\mathrm{Re}\left\{\sum_{n \in N} B_{en} \mathbf{V}_n e^{jn\omega_1 t}\right\} \tag{2.152}$$

Reactive current occurs when there is a phase shift between voltages and current harmonics (related to load susceptance), this current component does not participate in energy transfer from a source to the load.

$$\mathbf{i}_u = \sqrt{2}\mathrm{Re}\left\{\sum_{n \in N} A_n \mathbf{V}_n^{\#} e^{jn\omega_1 t}\right\} \tag{2.153}$$

Imbalance of the receiver causes the appearance of unbalanced current, it occurs only in three-phase systems, like in the previous component which also does not participate in energy transfer from a source to the load.

The theory proposed by Czarnecki was introduced to allocate the constituent phenomena in electrical systems, so that they can be correctly interpreted. However, its practical use in systems to improve power quality is difficult to achieve. Already the author in one of his publications shows that the compensator using this theory would be very complicated. Also applied is the theory of assumption of the orthogonality components. This theory used in other theories, was in the past often criticised by many scientists.

References

1. IEEE Trial (2000) Use standard definitions for the measurement of electric power quantities under sinusoidal, nonsinusoidal, balanced or unbalanced conditions. IEEE std, pp 1459–2000
2. IEEE Working Group on Nonsinusoidal Situations (1996) Practical definitions for powers in systems with nonsinusoidal waveforms and unbalanced loads: a discussion. IEEE Trans Power Delivery 2(1):79–101
3. IEEE (1992) The new IEEE standard dictionary of electrical and electronics terms. IEEE, New York
4. International Electrotechnical Commission (IEC) (1979) Technical Committee. No. 25. Working group 7 report: Reactive power and distortion power
5. Fryze S (1966) Selected problems of basics of electrical engineering (in Polish), PWN, Warszawa–Wrocław
6. A.I.E.E. Committee (1920) Apparent power in three-phase systems. Trans AIEE 39:1450
7. Buchholz F (1922) Die Drehstrom-Scheinsleistung bei ungleichmassiger Belastung der drei Zweige. Licht und Kraft 2:9–11
8. Emanuel AE (1998) The Buchholz-Goodhue apparent power definition: The practical approach for nonsinusoidal and unbalanced systems. IEEE Trans Power Delivery 13(2): 344–350
9. Budeanu CI (1928) Puissance reactives et fictives. RGE T. XXIII:762–773
10. Krogeris A, Rasevics K, Sinka J, Treimanis E (1993) Power in circuits of nonsinusoidal currents and voltages. Institute of Physical Energetics Latvian Academy of Sciences, Riga
11. Fryze S (1931) Active, reactive and apparent power in circuits with nonsinusoidal voltages and currents. (in polish) Przegląd Elektrotechniczny, 7:193–203, 8:225–234, 22: 673–676
12. Czarnecki LS (1987) What is wrong with the Budeanu concept of reactive and distortion powers and why is should be abandoned. IEEE Trans Instrumentations Meas 36(3):834–837
13. Emanuel AE (1993) Apparent and reactive powers in three-phase systems: in search of a physical meaning and better resolution. ETEP 3(1):7–14
14. Filipski P (1993) Apparent power a misleading quantity in the nonsinusoidal power theory: are all nonsinusoidal power theories doomed to fail? ETEP 3(1):21–26
15. Filipski P, Labaj P (1992) Evaluation of reactive power meters in the presence of high harmonic distortion. IEEE Trans Power Delivery 7(4):1793–1799
16. Filipski P (1980) A new approach to reactive current and reactive power measurement in nonsinusoidal systems. IEEE Trans Instrumentations Meas IM-29:423–426
17. Pasko M (1994) Selection of compensators to optimize the working conditions of single-phase and multi-phase voltage sources with periodical nonsinusoidal waveforms (in polish) Scientific papers of the Silesian University of Technology "Elektryka", vol 135
18. Peng FZ, Tolbert LM (2000) Compensation of non-active current in power systems, definitions from compensation standpoint. In: Proceedings 2000 IEEE PES summer meeting, pp 1–5
19. Piróg S (2001) Active power filters and STATCOM fundamental harmonic of reactive current compensator controlled in dq rotating coordinates (in polish). JUEE 7(1):91–105
20. Supronowicz H, Olszewski A (2001) Series active compensation (in polish). JUEE 7(1): 77–90
21. Shepherd W, Zakikhani P(1972) Suggested definition of reactive power for nonsinusoidal systems. Proc Inst Elec Eng 119(9):1361–1362, 120(7):705–706
22. Pasko M, Siwczyński M, Walczak J (1995) Why disappoint orthogonal current distribution for systems with nonsinusoidal waveforms? (in polish). In: EPN conference, pp 106–113
23. Sharon D (1973) Reactive power definitions and power factor improvement in nonlinear systems. Proc Inst Elec Eng 120(6):704–706
24. Emanuel AE (1974) Suggested definition of reactive power nonsinusoidal systems. Proc IEEE 121(7):705–706

25. Kusters NL, Moore WJM (1980) On the definition of reactive power under nonsinusoidal conditions. IEEE Trans Power Apparatus Syst 99:1845–1854
26. Czarnecki LS (1984) Interpretation, identification and modification of the energy properties of single-phase circuits with nonsinusoidal waveforms (in Polish) Scientific papers of the Silesian University of Technology "Elektryka", vol 91
27. Czarnecki LS (1991) Scattered and reactive current, voltage, and power in circuits with nonsinusoidal waveforms and their compensation. IEEE Trans Instrumentations Meas 40(3): 563–567
28. Pasko M (1992) Notes on scattering current compensation using LC systems for linear system powered by ideal and non-ideal distorted voltage sources (in polish), In: XV IC-SPETO conference, pp 519–525
29. Walczak J (1990) The distribution of the one-port load current supplied by nonideal and nonsinusoidal voltage source (in polish). In: IC-SPETO conference, pp 295–308
30. Pasko M, Walczak J (1996) Optimization of energy-quality properties of electrical circuits with periodical nonsinusoidal waveforms (in polish) Scientific papers of the Silesian University of Technology "Elektryka", vol 150
31. Siwczyński M (1995) Optimization methods in the power theory of electrical circuits (in polish). Cracow University of Technology Press, Poland
32. Walczak J (1992) Optimization of energy-quality properties of electrical circuits in Hilbert spaces (in polish) scientific papers of the Silesian University of Technology "Elektryka", vol 125
33. Pasko M (1995) Modification of three-phase systems with nonsinusoidal waveforms for optimization of source current shape. Archiwum Elektrotechniki XLIV(1):69–82
34. Akagi H, Kanazawa Y, Nabae A (1984) Instantaneous reactive power compensators comprising switching devices without energy storage components. IEEE Trans Ind App 1A-20(3):625–630
35. Czarnecki LS (2004) On some misinterpretations of the instantaneous reactive power p–q theory. IEEE Trans Power Electron 19(3):828–836
36. Pasko M, Maciążek M (2006) P-Q instantaneous power theory–a correct theory or useful algorithm for switched compensator control (in polish). Przegląd Elektrotechniczny 6:34–39
37. Komatsu Y, Kawabata T (1995) A control method of active power filter where system voltage contains negative-phase-sequence component or zero-phase-sequence component. PEDS Conf 2:583–586
38. Komatsu Y, Kawabata T (1999) A control method for the active power filter in unsymetrical voltage systems. Int J Electron 86(10):1249–1260
39. Komatsu Y, Kawabata T (1995) A control method for the active power filter in unsymetrical voltage systems. In: Proceedings of EPE'95, pp 1.904–1.907
40. Bhattacharya S, Divan DM, Banerjee B (1991) Synchronous frame harmonic isolator using active series filter. In: EPE 91 conference proceedings, vol 3, pp 3030–3035
41. Strzelecki R (1997) Application of the instantaneous power theory to active power filters control (in polish). JUEE 3:65–74
42. Czarnecki LS (2008) Currents' physical components (CPC): fundamental of power theory. Przegląd Elektrotechniczny (Proceedings of Electrical Engineering) R84, no. 6, pp 28–37

Chapter 3
Power Theories Applications to Control Active Compensators

Marcin Maciążek

Abstract One of the methods for high harmonic reduction, reactive power compensation or symmetrization in three-phase system is using active power filter (APF). APF connected to power system, depending on control strategy and configuration, can realise all of these functions or only selected ones. For example, parallel APF systems allow to compensate fundamental frequency of reactive component of load current, symmetrize load connected to network output terminals, filter higher harmonics of currents regardless of network impedance and with efficiency which is not accessible in the case of passive *LC* filters. This chapter concerns control algorithms for APFs. Theoretical considerations have been supplemented by simulation and experimental results.

3.1 Control Strategies for Active Compensators

Elimination of higher harmonics from current and voltage waveforms may be achieved by using compensators connected between source and load. So far mostly passive compensators were used, due to their simple design [1–7]. They are constructed as passive resonance filters matched to dominant harmonic frequency. Harmonic damping in these filters is strongly dependent on supply line parameters and is also limited by *Q* factor of used choke coils. These limitations are absent in

M. Maciążek (✉)
Silesian University of Technology, 2 Akademicka Street,
44-100 Gliwice, Poland
e-mail: marcin.maciazek@polsl.pl

G. Benysek and M. Pasko (eds.), *Power Theories for Improved Power Quality*,
Power Systems, DOI: 10.1007/978-1-4471-2786-4_3,

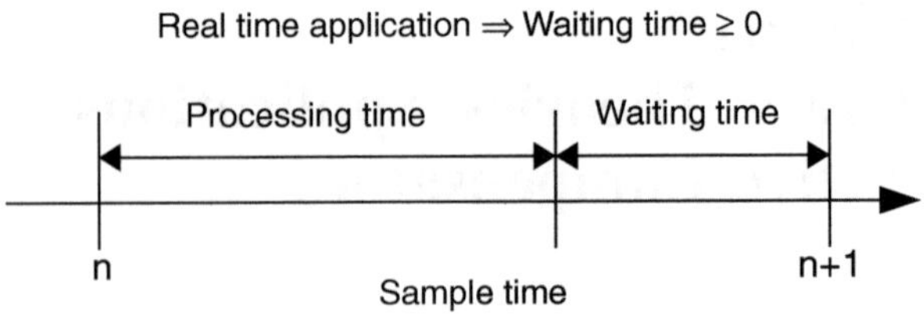

Fig. 3.1 Illustration of "real time" concept

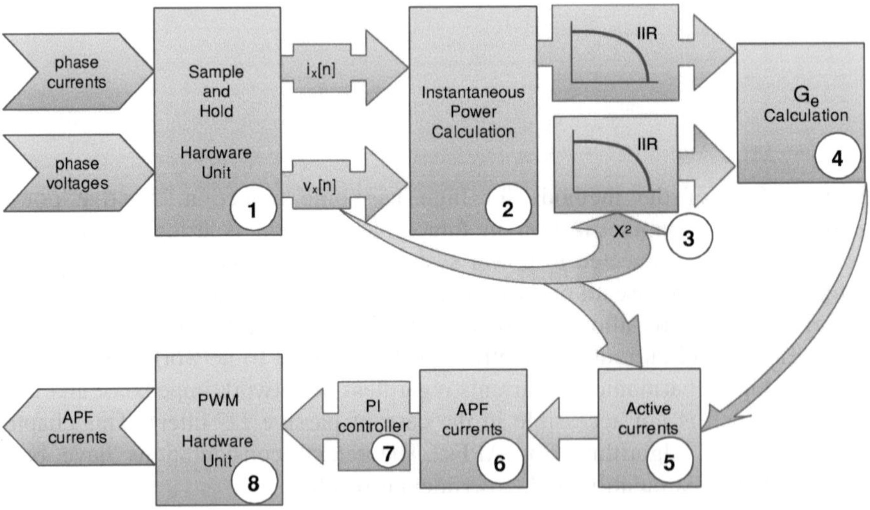

Fig. 3.2 Block diagram of control algorithm using Fryze power theory

case of active switching compensators, i.e. Active Power Filters (APF). Apart from eliminating harmonics and depending on control algorithm and structure used, these compensators make possible other compensating actions such as symmetrisation of source or load [8]. For these reasons APFs constitute the subject of numerous publications related to their design and operating principles [9–18].

Sinusoidal waveform is considered optimal in power systems and APF compensates instantaneous deviations of current/voltage waveform from this ideal. In general APF may be described as power electronic source of additional current (or voltage), connected in parallel (or in series) to the load. Sum of filter's current (voltage) and supply line current (voltage) will result in source's current (or load's voltage) being sinusoidal (in ideal case). All undesirable components of voltage or current close in load-APF circuit and do not have loading impact on the supply source. Bridge inverters using IGBT or MOSFET are used to form current/voltage source (mostly VSIs are used, with energy storage provided by capacitor, rarely current inverters with inductive energy storage). The inverters are controlled in such a way that output waveforms follow-up standard (model) waveforms.

APF are designed to operate in different supply systems such as:

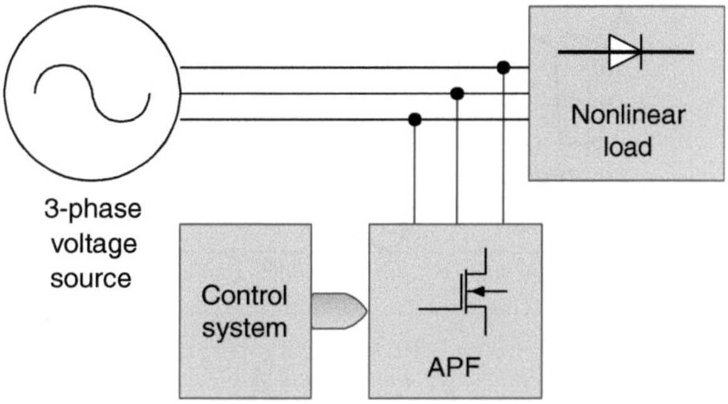

Fig. 3.3 Block diagram of three-phase system with nonlinear load and APF

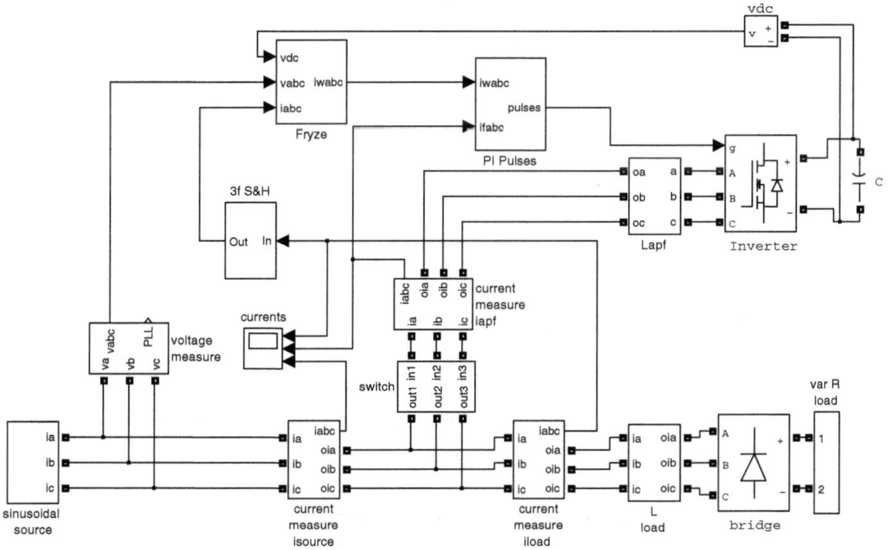

Fig. 3.4 Block diagram of model elaborated in Matlab/Simulink environment

- one-phase;
- three-phase three-wire (this is most common system);
- three-phase four-wire (system which makes possible compensation of current in neutral wire).

This classification is necessary to apply different control methods; these are thoroughly described in references [19, 20]. Other classification arises from the way APF is connected into source-load circuit, filtration properties of the filter result from connection type.

- Parallel connection—APF constitutes an additional current source. Usually power electronic voltage source inverters (VSI) are applied with inductors used in shaping follow-up current waveforms. This circuit makes it possible to:

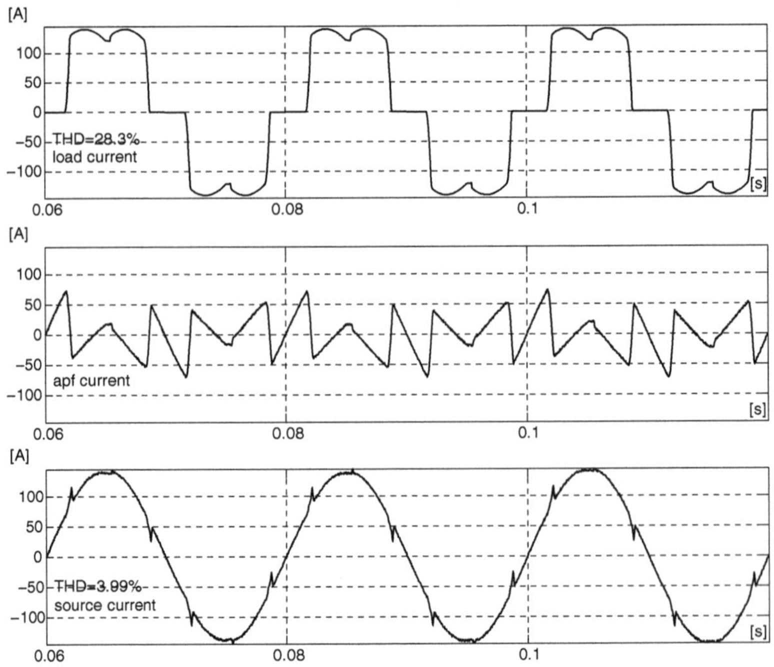

Fig. 3.5 Simulation results for phase *a*

- eliminate higher harmonics of source current at the level impossible to be attained by resonance filters;
- load symmetrisation;
- compensation of reactive power of fundamental component.

- Series connection—APF is connected in series with source-load circuit and is used to eliminate instantaneous voltage variations. Compensating voltage is introduced by a transformer. In design stage it must be taken into account that one of the windings will conduct load current with possible high values. This series circuit makes it possible to:
  - eliminate voltage's higher harmonics;
  - eliminate flickering (i.e. light flickering);
  - symmetrise supply (network) voltage.
- Series—parallel connection (so-called unified power flow controllers systems–UPFC), which exhibit properties peculiar to both series and parallel systems. What is important here is the sequence of connections of different sections (first the series circuit and then parallel circuit). This ensures correct operation of the circuit since voltage at parallel block terminals and current flowing through series block are almost sinusoidal (voltage harmonics are eliminated by the series section and current harmonics by the parallel section of the UPFC system).

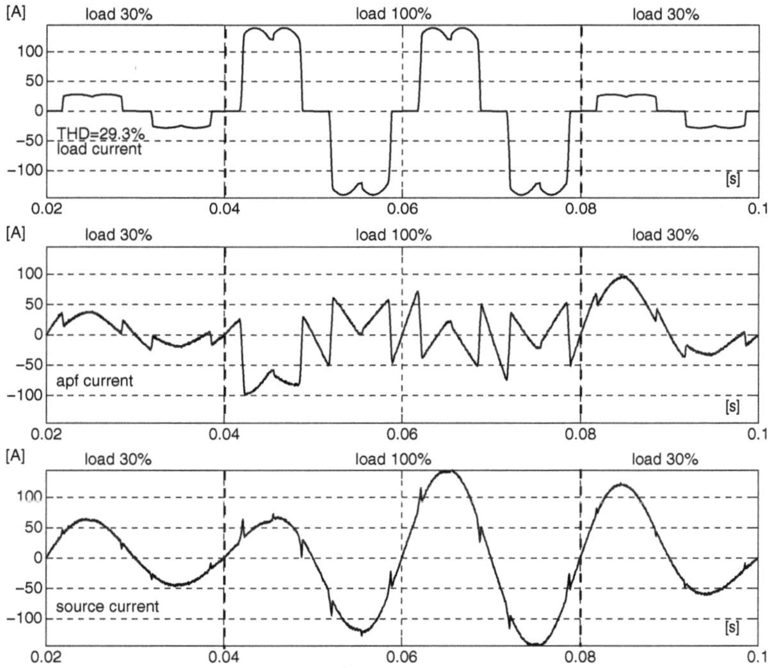

Fig. 3.6 Simulation results for phase *a*

If blocks changed places, higher harmonics would appear in waveforms and this would seriously complicate control of the whole system. These UPFC systems demonstrate features of series and parallel circuits both;

- Hybrid connection, which is made up of passive filter connected with APF. Passive filter may be connected to APF either in series or in parallel. Passive filter lightens the load of filter's active portion, reducing apparent power of this section by 60% or more. Voltage present at terminals of power electronic valves (MOSFET of IGBT transistors) is much lower than in traditional circuits. This lowers construction costs and makes it possible to design and build filters with higher apparent power. Uniting passive and active filter characteristics in one system results in good filtration properties with many faults eliminated at the same time (e.g. the impact of network impedance in passive filter is significantly lowered).

Active compensator requires different control algorithms depending on its connection and type of network where it operates. The basic goal of control algorithm is to determine correctly reference compensation currents and to introduce them into supply network. Due to classification given above, large number of different control cases is feasible; it is not therefore possible to present all of them in the current monograph. We have decided to elaborate on the most common supply system that is, three-phase three-wire system with APF connected

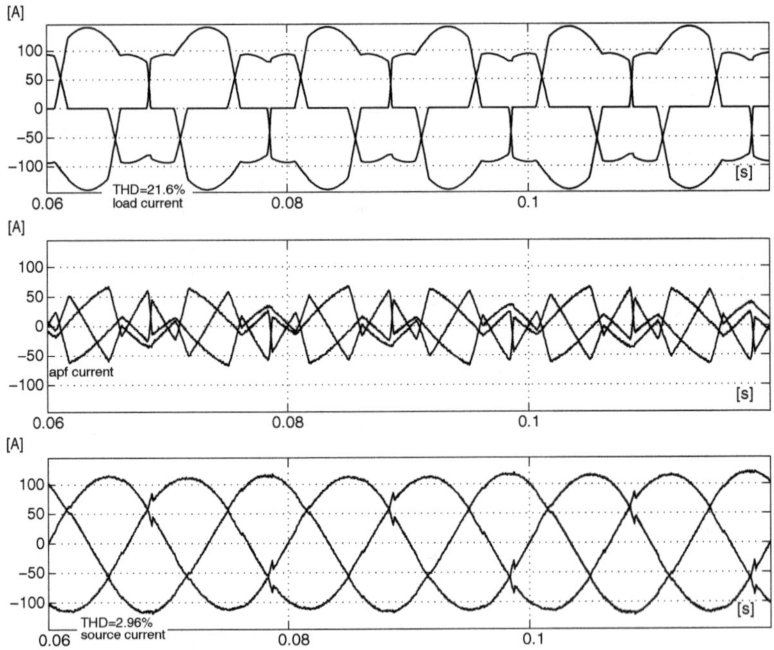

Fig. 3.7 Simulation results, THD coefficients relate to phase *a*

in parallel. Reference compensation currents will be determined using the most popular power theories demonstrated previously in Chap. 2.

The proposed control algorithms have been subjected to simulations under two different supply conditions, i.e.:

- sinusoidal source with zero internal impedance;
- non-sinusoidal source with non-zero internal impedance; and two different loads:
- symmetrical load;
- non-symmetrical load.

Results of simulations will be presented later in the chapter.

3.1.1 Control Strategy Based on Fryze Theory

Control algorithm using Fryze's theory [21] to determine reference compensation currents operates in time domain and calculations are conducted in real time. The "real time" concept is irrevocably related to the given problem and used hardware. It is understood that if $x[n]$ denotes input signal $x(t)$ data and $y[n]$ denotes output data, then processing of this signal is run in real time if algorithm is completed

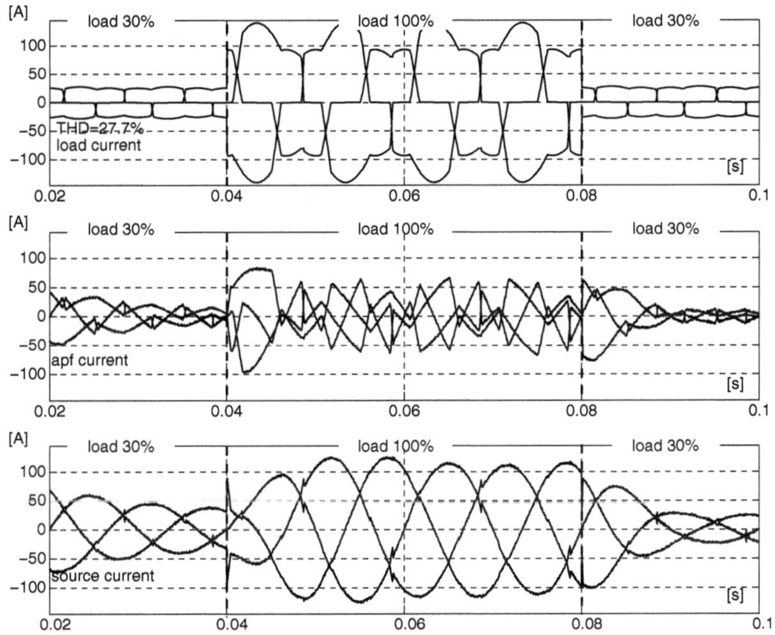

Fig. 3.8 Simulation results, THD coefficients relate to phase *a*

before next sample ($x[n + 1]$) appears in circuit input. The idea is shown in Fig. 3.1.

After analysing theory proposed by Fryze [21], control algorithm shown in Fig. 3.2 has been developed.

In case of exemplary three-phase circuit shown in Fig. 3.3, the whole of control algorithm is run in interrupt handling procedure generated in hardware PWM block of the processor. Knowing acquired instantaneous values of phase currents and load voltages (the acquisition is achieved by hardware module of the processor, block {1}), instantaneous power value is calculated in accordance with formula:

$$p[n] = i_a[n]v_a[n] + i_b[n]v_b[n] + i_c[n]v_c[n] \tag{3.1}$$

Later, average value of instantaneous power (i.e. load's active power—block {2}) is calculated using low-pass filter IIR. This procedure corresponds to equation:

$$P = \frac{1}{T}\int_0^T v(t)i(t)\mathrm{d}t \tag{3.2}$$

or:

$$P[n] = \overline{p[n]} = b_0p[n] + b_1p[n-1] + b_2p[n-2] - a_1P[n-1] - a_2P[n-2] \tag{3.3}$$

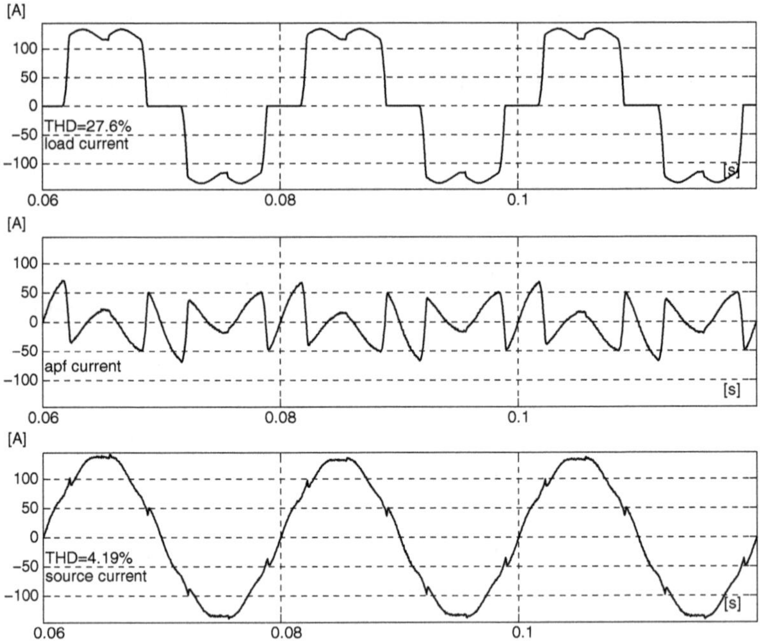

Fig. 3.9 Simulation results, phase *a*

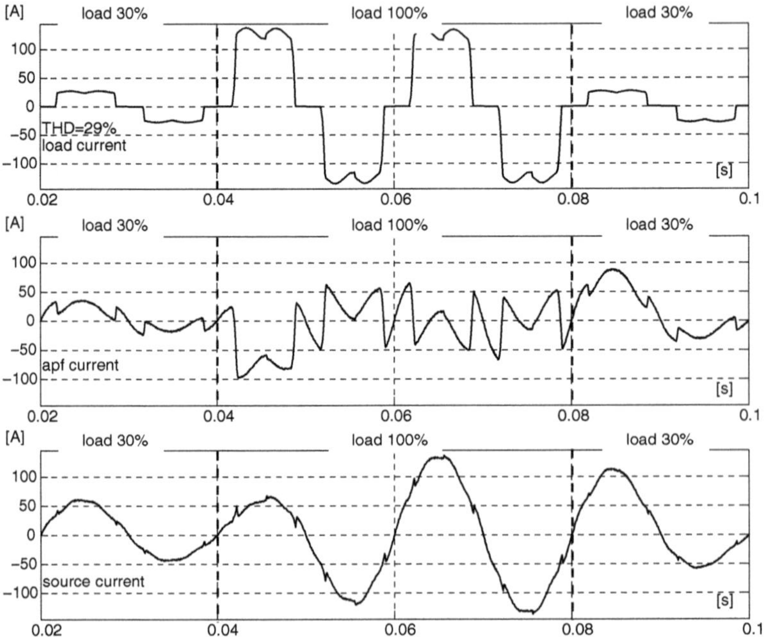

Fig. 3.10 Simulation results, phase *a*

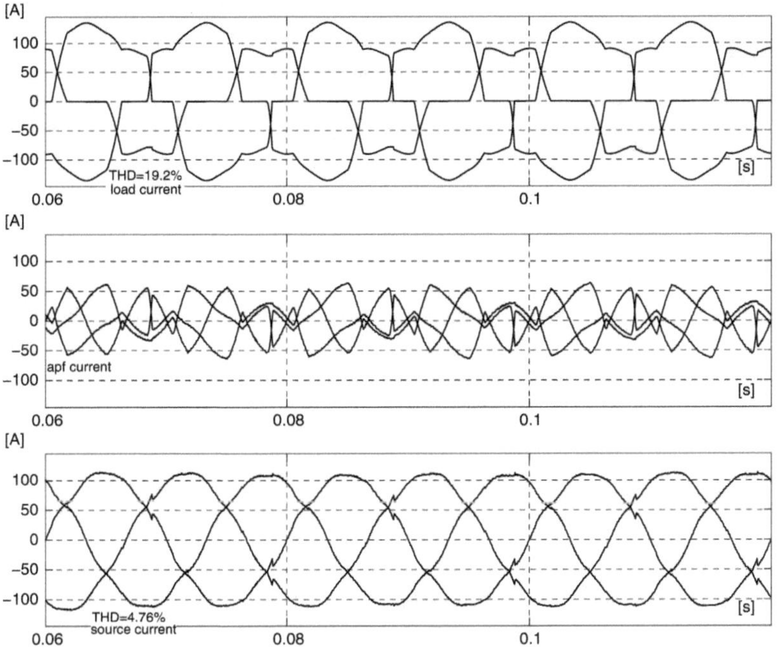

Fig. 3.11 Simulation results, THD coefficients relate to phase *a*

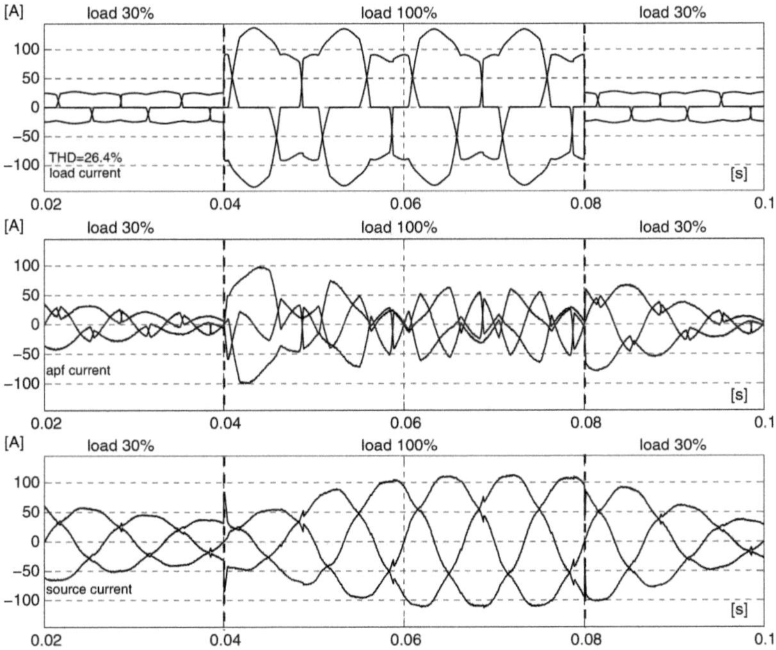

Fig. 3.12 Simulation results, THD coefficients relate to phase *a*

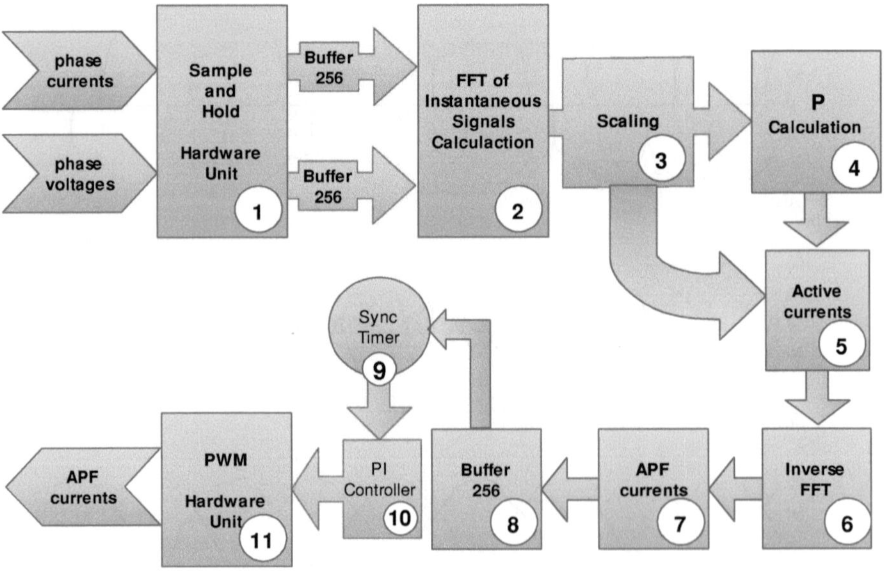

Fig. 3.13 Block diagram of control algorithm using Budeanu theory

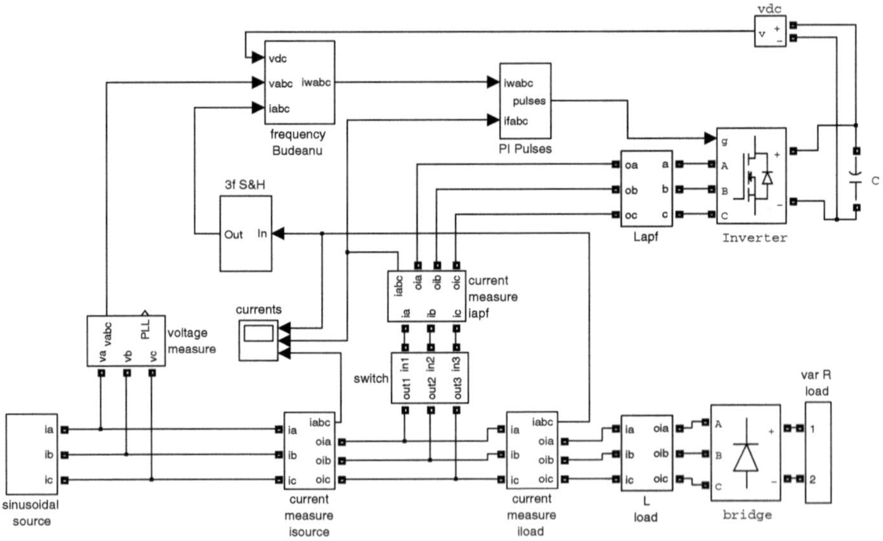

Fig. 3.14 Block diagram of model implemented in Matlab/Simulink environment

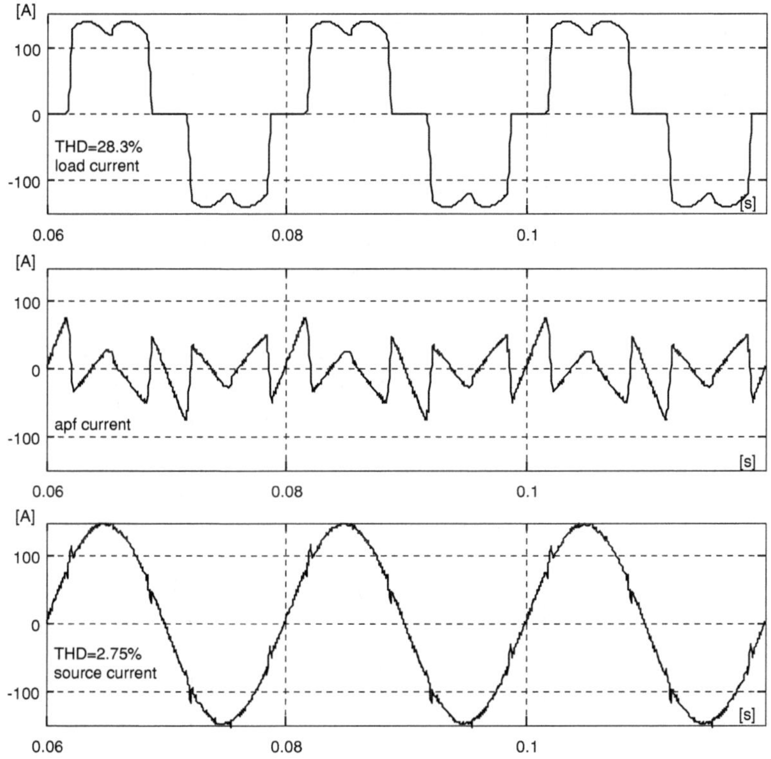

Fig. 3.15 Simulation results for phase *a*

where:
b_x, a_x—digital filter coefficients.
Block {3} is used to calculate average values of phase voltages squared instantaneous values. Again, low-pass IIR filters are used here. This corresponds to Eq. 3.4 in a way similar to Eq. 3.3:

$$\overline{v_x^2[n]} = \overline{v_x[n] \times v_x[n]}; \quad x = \{a, b, c\} \tag{3.4}$$

Determined values of active power (block {2}) and average values of squared phase voltages (block {3}) make it possible to determine equivalent load conductance (block {4}):

$${}_eG[n] = \frac{P[n]}{\overline{v_a^2[n]} + \overline{v_b^2[n]} + \overline{v_c^2[n]}} \tag{3.5}$$

In next step active currents sample values are determined (block {5}):

$$i_x^a[n] = {}_eG[n] v_x[n]; \quad x = \{a, b, c\} \tag{3.6}$$

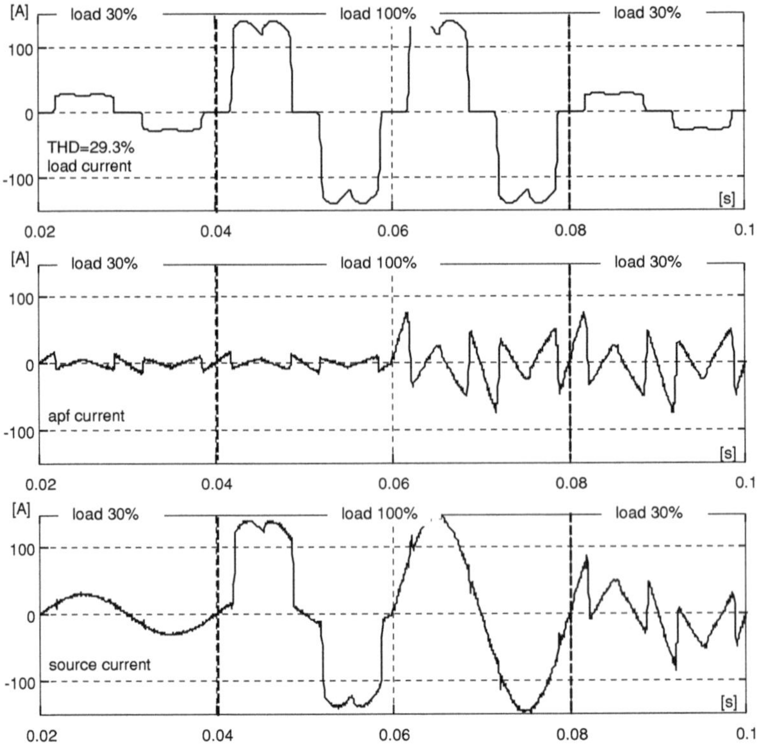

Fig. 3.16 Simulation results for phase *a*

According to Fryze, active current is the current of minimum RMS-value ensuring required active power of the load. Therefore, if we assume that source current should be calculated in accordance with above statement, the reference compensation APF current may be determined from equation given below (block {6}):

$$i_{fx}[n] = i_x[n] - i_x^a[n]; \quad x = \{a, b, c\} \tag{3.7}$$

Values determined in accordance with procedures described above are then transferred to controller input. Depending on selected design, this may be for example either PI controller or predictive controller (block {7}). Controller output is connected to the hardware block, which generates PWM signal (block {8}).

3.1.1.1 Model of Source-APF-Load Circuit in Matlab/Simulink Environment

Circuit shown in Fig. 3.3 has been simulated with the help of Matlab/Simulink programming environment. This model is shown in Fig. 3.4. It has been developed in such a way that the operation of microprocessor system is emulated as faithfully

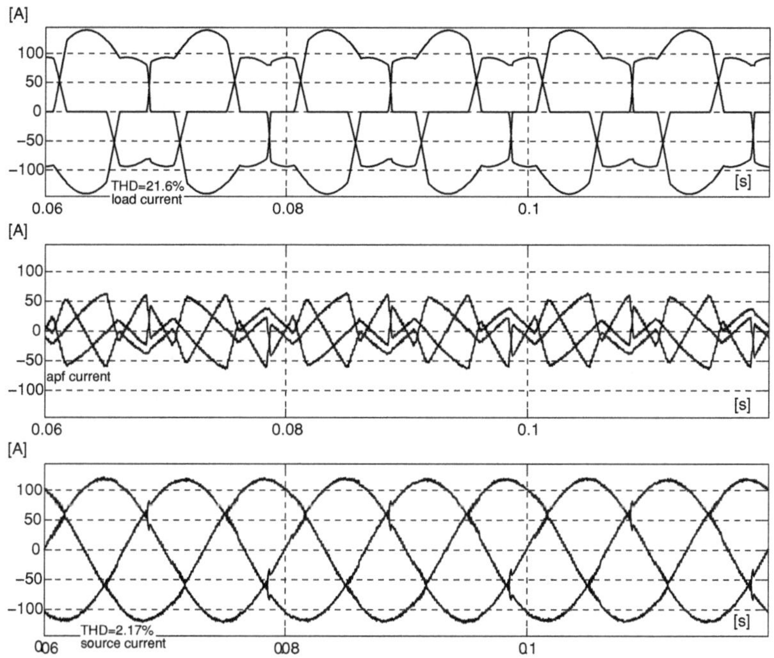

Fig. 3.17 Simulation results; THD coefficients relate to phase *a*

as possible. For instance, delays occurring in measurement and processing circuits have been taken into account by using S & H circuits.

3.1.1.2 Simulation Results

Circuit shown in Fig. 3.4 has been subjected to analysis under different operating conditions defined in the initial section of present chapter.

Results of simulation (steady state) in case of sinusoidal source with zero internal impedance and symmetrical load are shown in Fig. 3.5.

Simulation results for transient state in case of sinusoidal source with zero internal impedance and symmetrical load are shown in Fig. 3.6.

Results of simulation (steady state) in case of sinusoidal source with zero internal impedance and non-symmetrical load are shown in Fig. 3.7.

Results of simulation for transient state in case of sinusoidal source with zero internal impedance and non-symmetrical load are shown in Fig. 3.8.

Simulation results for steady state in case of non-sinusoidal source with non-zero internal impedance and symmetrical load are shown in Fig. 3.9. It must be noted that for this type of source, source current components defined by Fryze theory are no longer orthogonal.

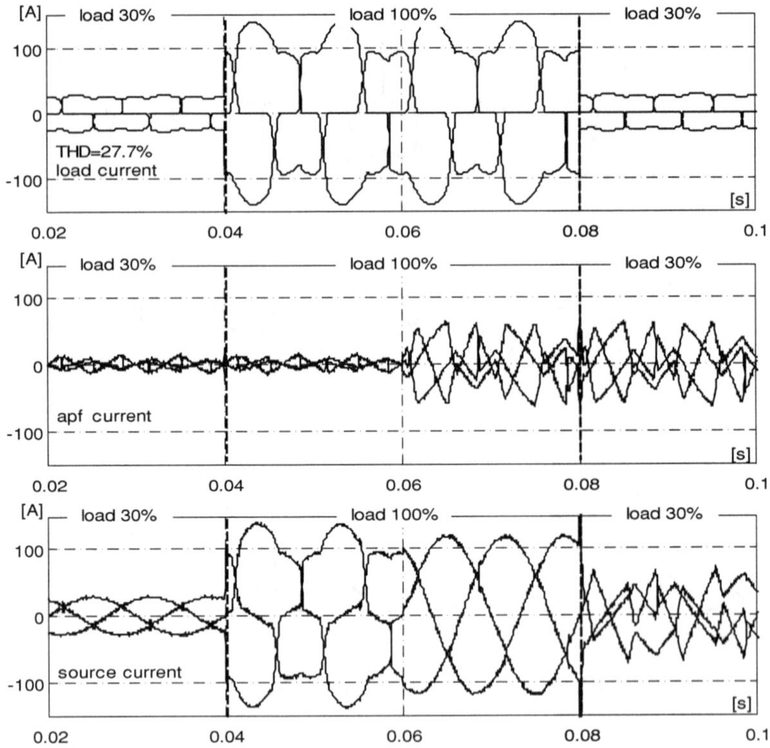

Fig. 3.18 Simulation results; THD coefficients relate to phase *a*

Simulation results for transient state in case of non-sinusoidal source with non-zero internal impedance and symmetrical load are shown in Fig. 3.10.

Simulation results for steady state in case of non-sinusoidal source with non-zero internal impedance and non-symmetrical load are shown in Fig. 3.11.

Simulation results for transient state in case of non-sinusoidal source with non-zero internal impedance and non-symmetrical load are shown in Fig. 3.12.

3.1.1.3 Conclusions

- Proposed algorithm fulfils correctly the desired elimination of unwanted components of source current. In case of non-symmetrical load and when compensation is used, the source current is symmetrical.
- Necessity of using several low-pass filters in the system has adversely affected dynamic properties. It is true that instantaneous power step changes ranging from 30 to 100%, which have been applied in simulations, are rarely met with in real life, still this kind of algorithm should be optimised with specific application

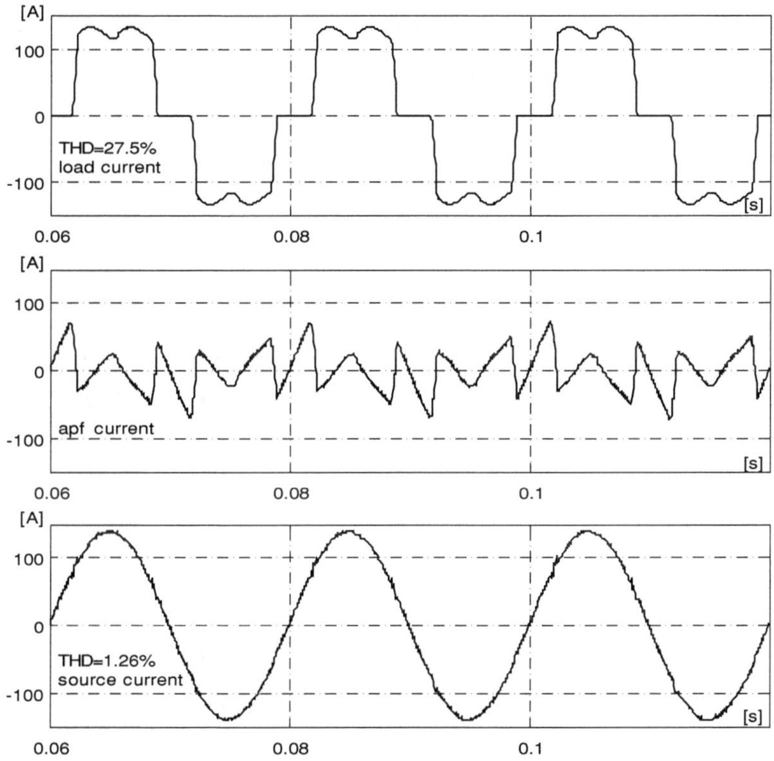

Fig. 3.19 Simulation results—phase *a*

in view. In such case, filter characteristics used should be modified in order to minimise time needed to stabilise waveform.

- Equation 3.6 shows that the active component mirrors the shape of supply voltage waveform. Therefore, in case of supplying the system from source with non-zero internal impedance, partially uncompensated harmonics of fifth and seventh order remain in the source current. This is due to deformation of the supply voltage, where these particular harmonics have also been present. Even though in this particular case the current components defined by Fryze are no longer orthogonal, source current resulting from compensation ensures that the active power transferred to the load does not change.

3.1.2 Control Strategy Based on Budeanu Theory

Control algorithm basing on Budeanu power theory adopts frequency domain for analysis of circuit state. This requires specific methods allowing for representation of phase voltages and currents in the form of Fourier series. Block diagram of control algorithm is shown in Fig. 3.13.

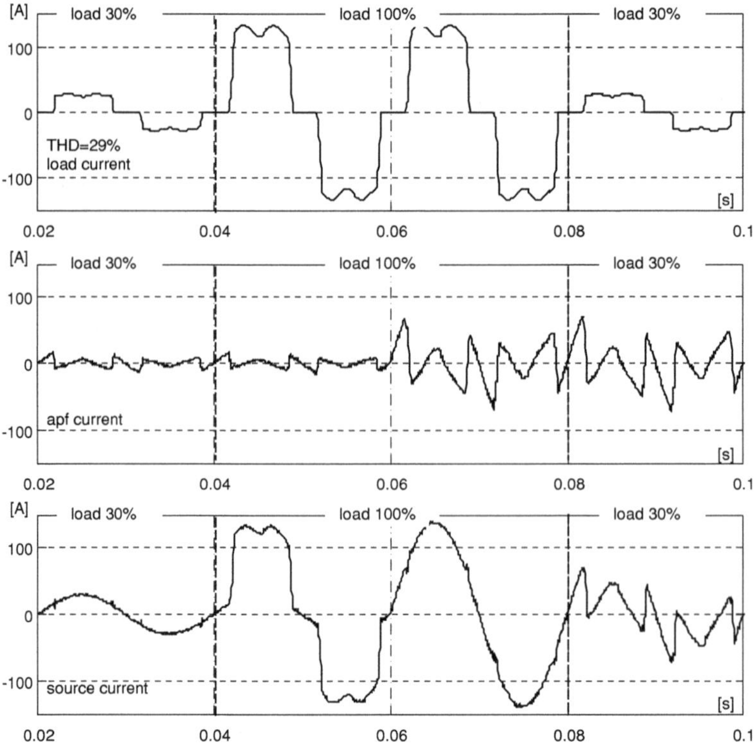

Fig. 3.20 Simulation results—phase *a*

Acquired instantaneous values of load phase voltages and currents (conducted by hardware module of the processor—block {1}) are stored in the buffer which allows for storage of 256 samples of each signal. Sample frequency of S&H circuit must be set so that acquisition time of 256 samples is equal to exactly 20 ms. This will ensure proper synchronisation of the algorithm with fundamental harmonic frequency. Hen buffer is full, sampled current signals are subjected to Fast Fourier Transform (FFT)—block {2}. Fourier transform makes it possible to represent time-varying signals in frequency domain. Discrete Fourier Transform (DFT) is used when values of signal being transformed are discrete, i.e. spaced evenly in time; interval between the signals is called sampling period. Computational complexity of DFT algorithm is the reason why FFT algorithm is usually used. This algorithm requires that the number of samples of the input signal is the multiple of two. In the presented example and taking into account the size of input buffer, FFT of 256 points is calculated. Correct representation of waveform component amplitudes (i.e. harmonics) requires additional scaling (this is done in block {3}). This means that amplitudes obtained by FFT must be divided by the data buffer length. Finally, as a result of these operations, phase voltages and currents may be expressed generally as:

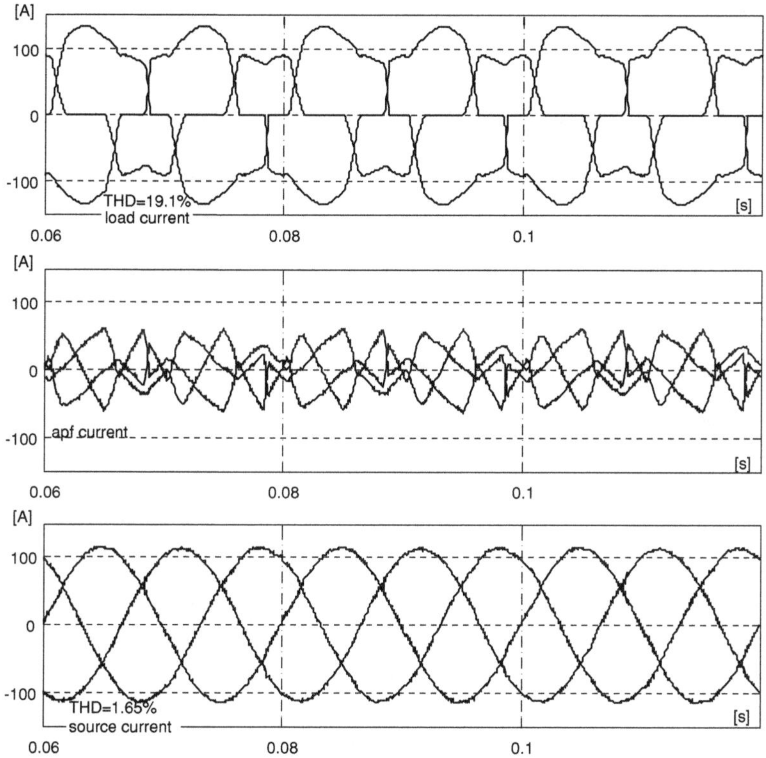

Fig. 3.21 Simulation results; THD coefficients relate to phase *a*

$$f(t) = F[0] + \sqrt{2}\mathrm{Re}\sum_{h=1}^{N} F[h]\exp(jh\omega t + \varphi[h]) \tag{3.8}$$

where:
$F[h]$—components of amplitude spectrum, $\varphi[h]$—components of phase spectrum.

Next, load active power may be calculated (block{4}):

$$P = \sum_{\mathrm{x=a,b,c}} V_x[0]I_x[0] + \sum_{\mathrm{x=a,b,c}} \sum_{h=1}^{N} V_x[h]I_x[h]\cos(\varphi_{vx}[h] - \varphi_{ix}[h]) \tag{3.9}$$

Then spectra of active currents should be determined (block {5}). If we assume that as a result of compensation we aim at obtaining sinusoidal current being in phase with voltage fundamental harmonic, then active current should be determined in accordance with formula (3.10):

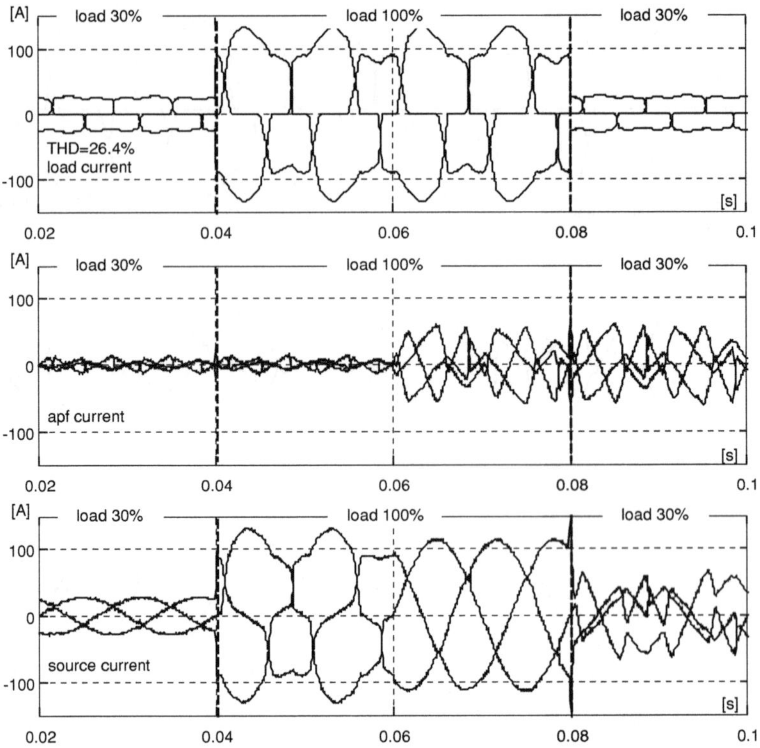

Fig. 3.22 Simulation results; THD coefficients relate to phase *a*

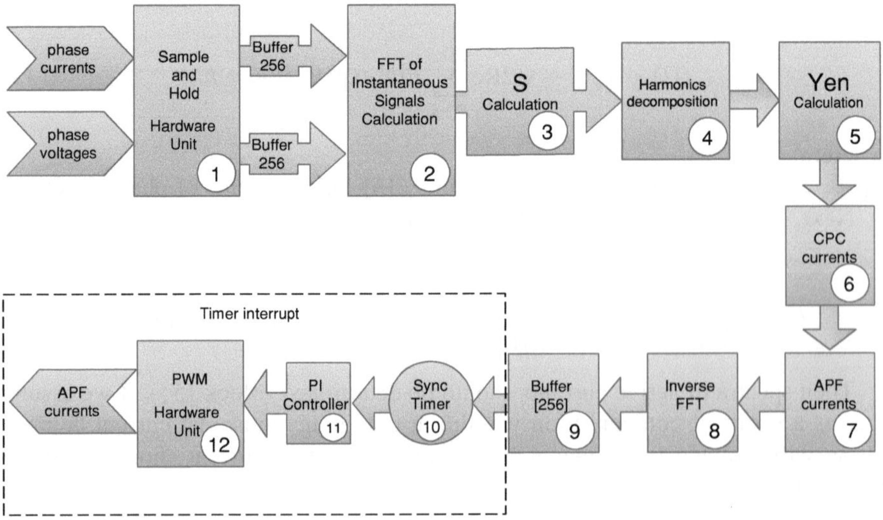

Fig. 3.23 Block diagram of control algorithm using CPC theory

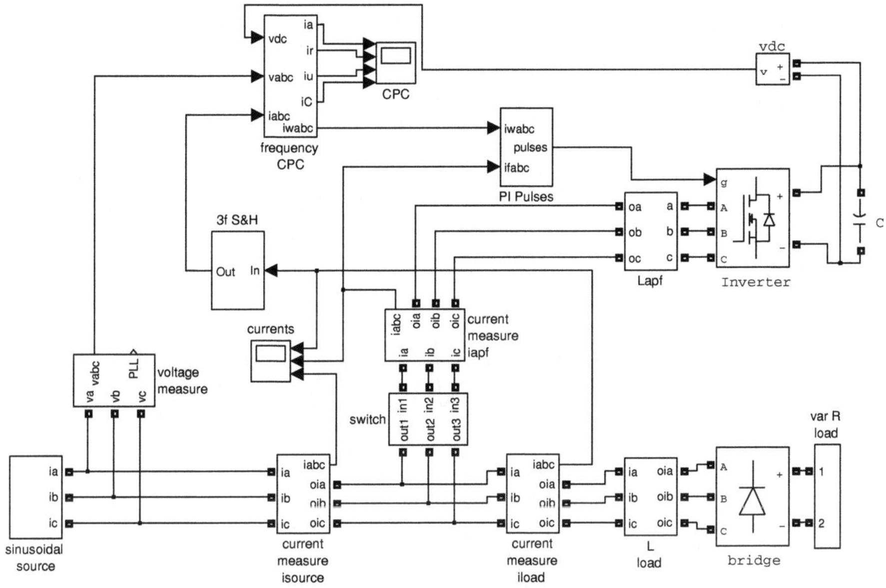

Fig. 3.24 Block diagram of model elaborated for Matlab/Simulink environment

$$I_{\mathrm{x}}^{a}[1] = \frac{2}{3}\frac{P}{V_{\mathrm{x}}[1]}; \quad x = a, b, c$$
$$\varphi_{\mathrm{x}}^{a}[1] = \varphi_{\mathrm{vx}}[1]; \quad x = a, b, c \tag{3.10}$$

The determined spectrum of active current is then subjected to Fourier's inverse transform (block{6}), and 256 samples of sinusoidal current ensuring required load active power are obtained as a result. The reference compensation currents are calculated by subtracting active current from load's phase current (block {7}).

$$i_{\mathrm{fx}}[n] = i_{\mathrm{x}}[n] - i_{\mathrm{x}}^{a}[n]; \quad n = \{1, 2, \ldots, 256\}; x = a, b, c \tag{3.11}$$

The samples of reference compensation currents obtained in this way will be placed in single output buffer (block {8}), from whence they will be acquired by synchronising circuit (block {9}), and through PI controller (block{10}) and hardware PWM block (block{11}) will be input to the system.

3.1.2.1 Model of Source-APF-Load Circuit in Matlab/Simulink Environment

Model circuit has been implemented in Matlab/Simulink environment—see Fig. 3.14. This model makes it possible to conduct analysis of control algorithm using Budeanu theory.

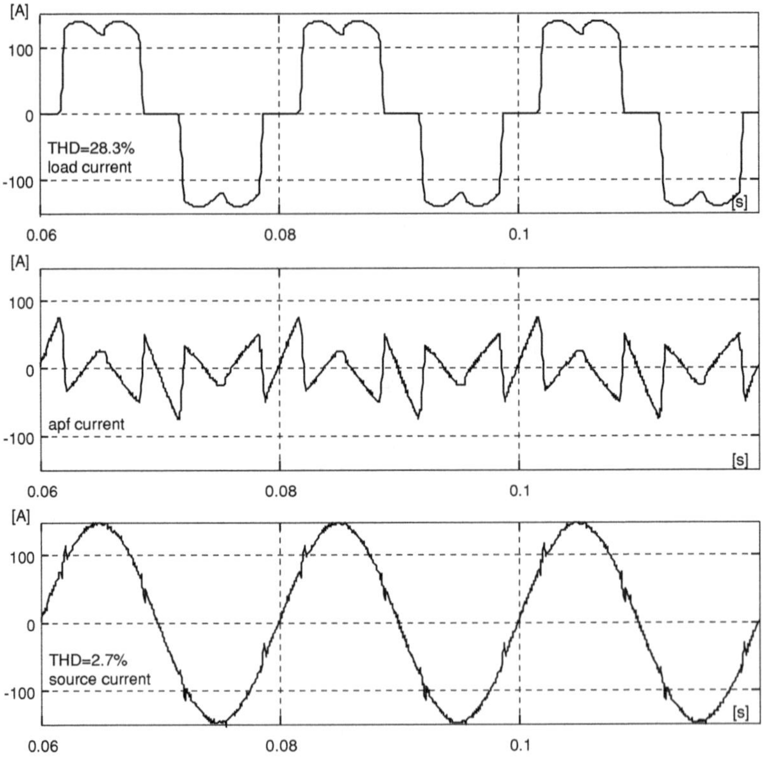

Fig. 3.25 Simulation results for phase *a*

3.1.2.2 Simulation Results

Circuit shown in Fig. 3.14 has been subjected to analysis under different operating conditions (same as in previous case). The simulation results of steady state in case of supply consisting of sinusoidal source and zero internal impedance and symmetrical load are shown in Fig. 3.15.

The simulation results of transient state in case of supply consisting of sinusoidal source and zero internal impedance and symmetrical load are shown in Fig. 3.16.

The simulation results of steady state in case of supply consisting of sinusoidal source with zero internal impedance and non-symmetrical load are shown in Fig. 3.17.

The simulation results of transient state in case of supply consisting of sinusoidal source with zero internal impedance and non-symmetrical load are shown in Fig. 3.18.

The simulation results of steady state in case of supply consisting of non-sinusoidal source with non-zero internal impedance and symmetrical load are shown in Fig. 3.19.

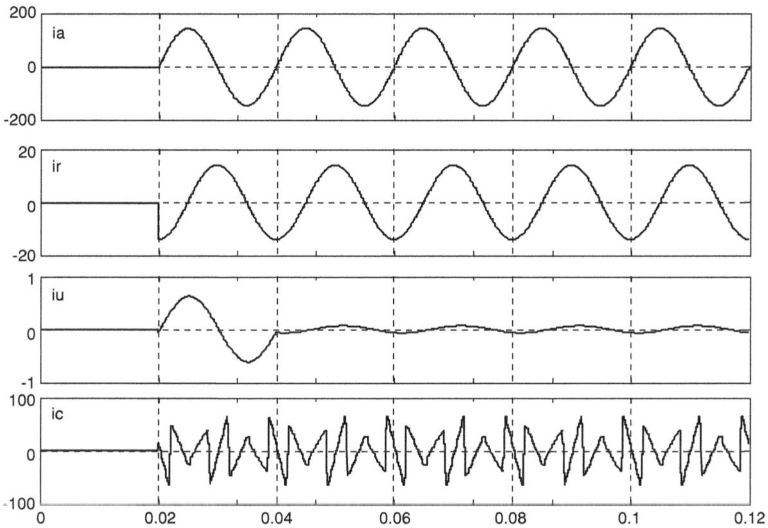

Fig. 3.26 CPC components, i.e. active, reactive, unbalance current and current generated for phase *a*

The simulation results of transient state in case of supply consisting of non-sinusoidal source with non-zero internal impedance and symmetrical load are shown in Fig. 3.20.

The simulation results of steady state in case of supply consisting of non-sinusoidal source with non-zero internal impedance and non-symmetrical load are shown in Fig. 3.21.

The simulation results of transient state in case of supply consisting of non-sinusoidal source with non-zero internal impedance and non-symmetrical load are shown in Fig. 3.22.

3.1.2.3 Conclusions

- Source current in steady states may be characterised by lower THD coefficient compared to results obtained with control strategies using time domain. This is due to possibility of simple shifting of reference compensation current waveform (being present in the buffer) and in this way deformations caused by delay of the reference are eliminated.
- Algorithm utilising frequency domain is theoretically more flexible when it comes to elimination of any chosen harmonics.
- In transient states source current waveforms are much more distorted than in time-based algorithms. This is caused by the fact that the correct signal reference is accessible during signal's second period at the earliest. In addition, the signal course in transient state is non-periodical and therefore should not be

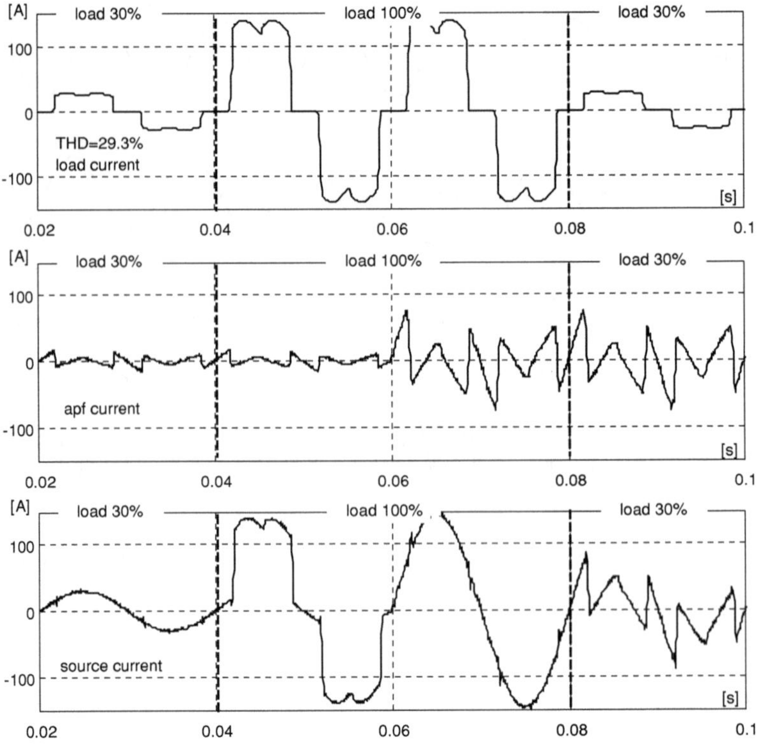

Fig. 3.27 Simulation results, phase *a*

analysed with the help of Fourier transform. This limitation is most serious since real voltage and current waveforms may usually be classed as nearly periodic waveforms.

3.1.3 Control Strategy Based on CPC Theory

Control algorithm based on CPC power theory uses frequency domain in order to analyse state of the system (same as in case of Budeanu theory). The block diagram of the control algorithm is shown in Fig. 3.23.

The acquired instantaneous values of phase currents and voltages supplying the load (implemented by processor's hardware module block {1}) are stored in buffer. This buffer allows for storage of 256 samples of each signal. When the buffer is full, current samples are subjected to FFT—block {2}. As a result, phase voltages and currents may be represented in the following form:

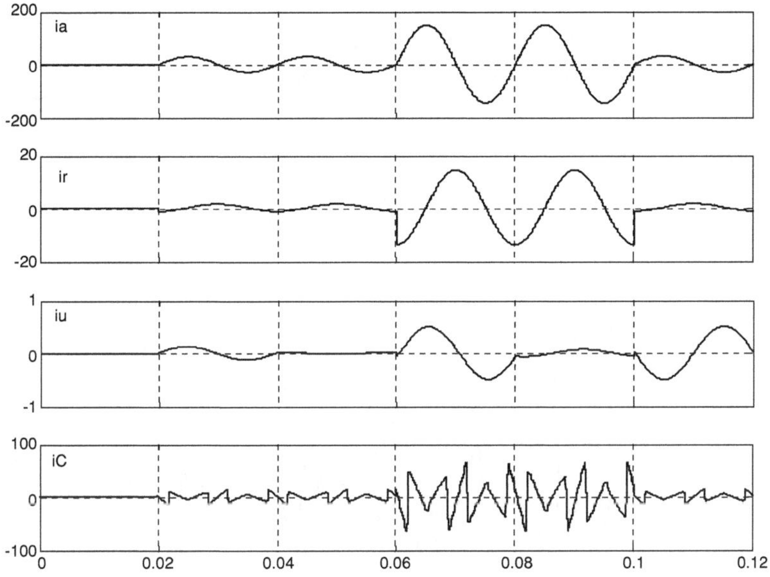

Fig. 3.28 CPC components, i.e. active, reactive, unbalance current and current generated for phase *a*

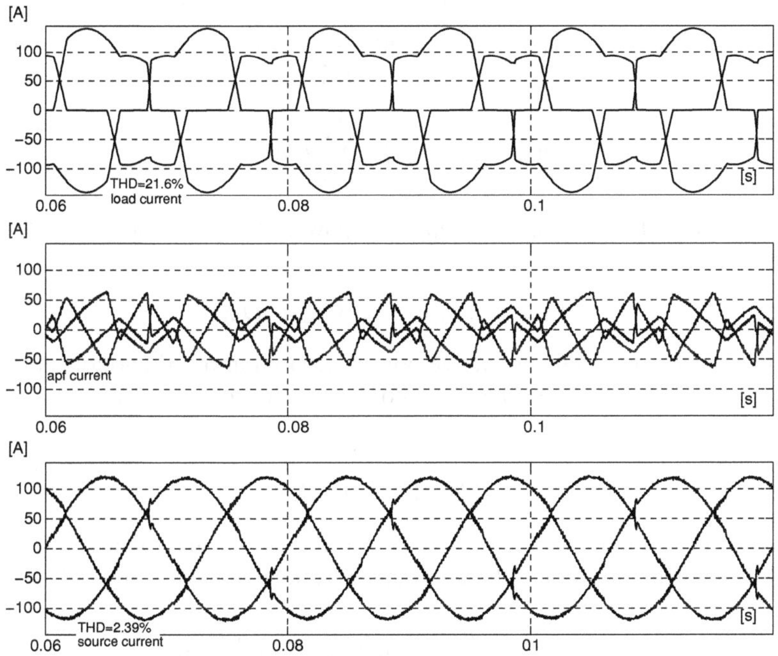

Fig. 3.29 Simulation results, THD coefficients relate to phase *a*

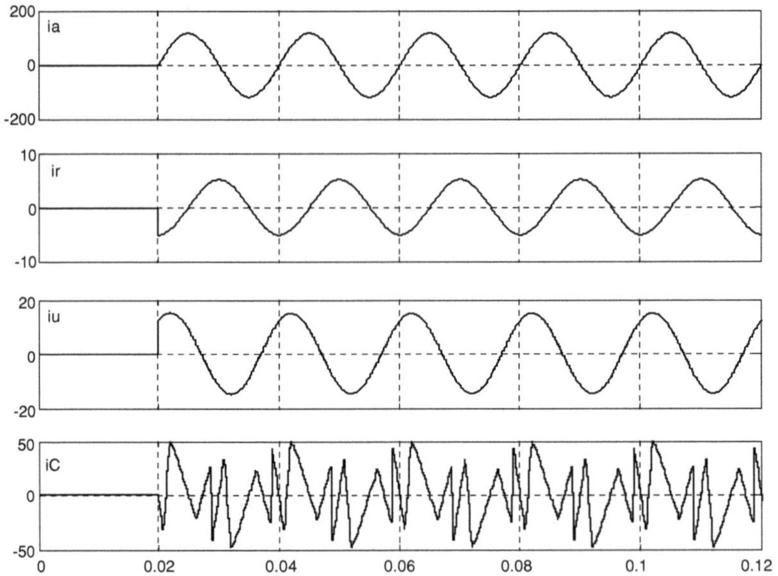

Fig. 3.30 CPC components, i.e. active, reactive, unbalance current and current generated for phase *a*

$$v(t) = V_0 + \sqrt{2}\mathrm{Re}\sum_{h=1}^{N} V[h]\mathrm{e}^{\mathrm{j}h\omega_1 t} \tag{3.12}$$

$$i(t) = I_0 + \sqrt{2}\mathrm{Re}\sum_{h=1}^{N} I[h]\mathrm{e}^{\mathrm{j}h\omega_1 t} \tag{3.13}$$

On the basis of these quantities, apparent power of each harmonic is determined (block {3}):

$$S[h] = V[h]I^*[h] = P[h] + jQ[h] \tag{3.14}$$

Again, based on calculated active power, separation of all harmonics into two subsets is achieved ({block (4)}) in accordance with the criterion:

$$\begin{aligned} P[h] &\geq 0, \quad h \in N_\mathrm{D} \\ P[h] &<; 0, \quad h \in N_\mathrm{C} \end{aligned} \tag{3.15}$$

In case of harmonics belonging to N_D set, the circuit may be investigated as a passively loaded circuit, and the admittances are calculated in block {5}:

$$Y_\mathrm{e}[h] = \frac{S^*[h]}{\|V[h]\|^2} = G_\mathrm{e}[h] + jB_e[h]\, h \in N_\mathrm{D}. \tag{3.16}$$

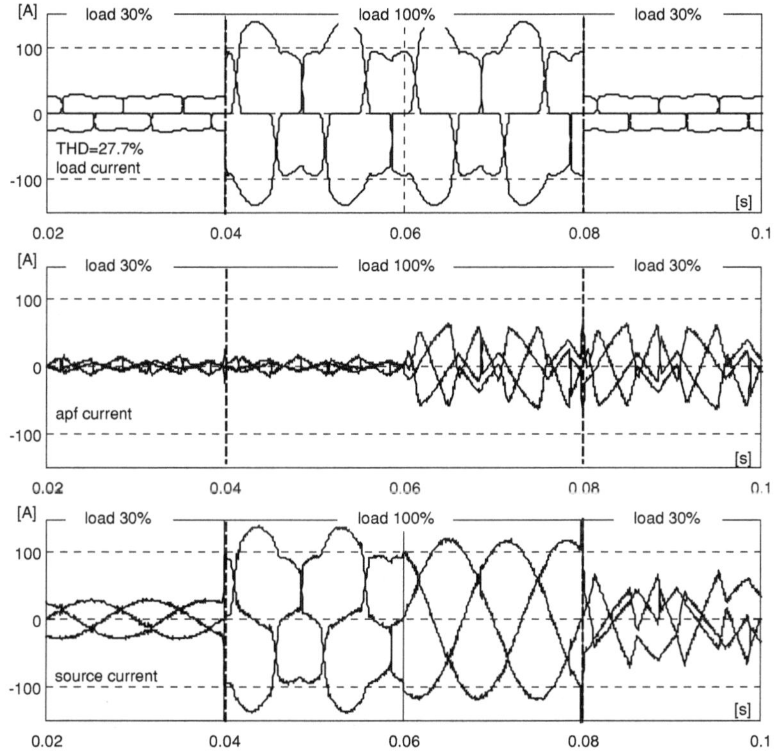

Fig. 3.31 Simulation results, THD coefficients relate to phase *a*

Using this data, the current components defined by CPC theory are determined in block {6}, i.e.:

- Active component

$$G_e = \frac{\sum\limits_{h \in N_D} P[h]}{\left\| \sum\limits_{h \in N_D} V[h] \right\|^2} \qquad (3.17)$$

$$I^a = G_e \sum_{h \in N_D} V[h]$$

- Reactive component

$$I^r[h] = B_e[h]V[h]; \quad h \in N_D \qquad (3.18)$$

- Scatter component

$$I^s[h] = (G_{eh}[h] - G_e)V[h]; \quad h \in N_D \qquad (3.19)$$

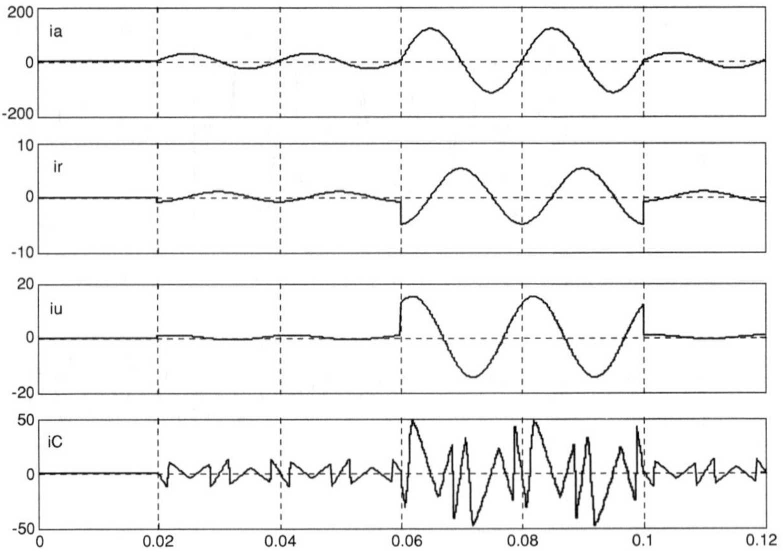

Fig. 3.32 CPC components, i.e. active, reactive, unbalance current and current generated for phase *a*

- Unbalance component

$$I^u[h] = A[h]V^{\#}[h]; \quad h \in N_{\mathrm{D}} \tag{3.20}$$

- Generated component

$$I^C[h] = I[h]; \quad h \in N_{\mathrm{C}} \tag{3.21}$$

Next, spectra of active currents must be determined (block {7}). If it is assumed that we aim at obtaining sinusoidal current being in phase with voltage's fundamental harmonic as a result of the compensation, standard compensating current must be calculated using the formula:

$$I[n] = I^r[n] + I^s[n] + I^u[n] + I^C[n] \tag{3.22}$$

The calculated current spectrum is then subjected to inverse Fourier transform (block {8}), and this results in 256 current samples. The current is sinusoidal and this ensures required load's active power. The samples of standard compensating currents obtained in this way will be placed in special output buffer (block {9}); from this buffer they will be acquired by synchronising circuit (block {10}). *PI* controller (block {11}) and hardware PWM block (block {12}) will be used to introduce them into the system.

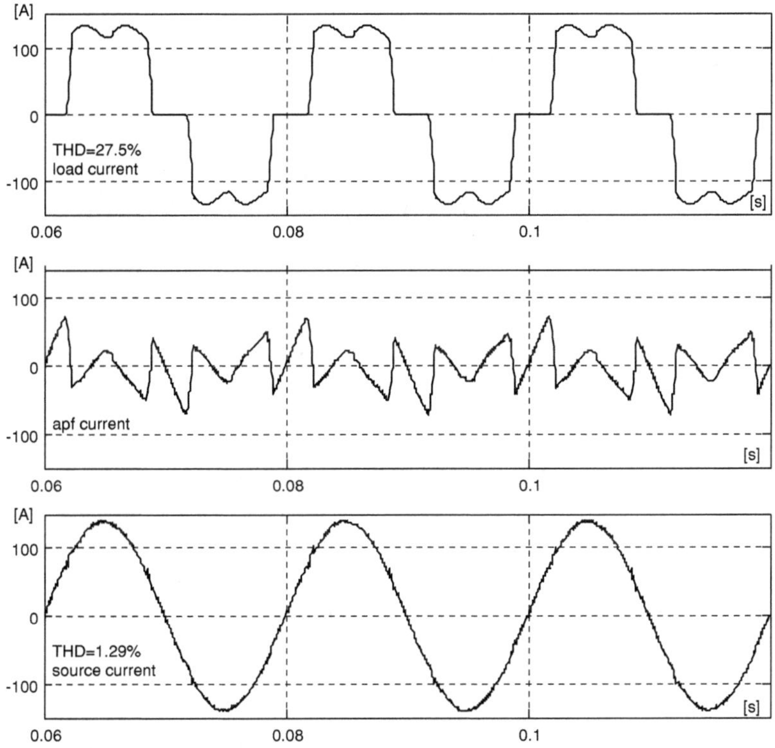

Fig. 3.33 Simulation results phase *a*

3.1.3.1 Model of Source-APF-Load System in Matlab/Simulink Environment

Analysis of control algorithm using CPC theory has been run with the help of circuit model designed purposely for Matlab/Simulink environment. This model is shown in Fig. 3.24.

3.1.3.2 Simulation Results

Circuit shown in Fig. 3.24 has be analysed under different operating conditions (same as in case of algorithms described previously). Simulation results for steady state in case of sinusoidal source with zero internal impedance and symmetrical load are shown in Fig. 3.25.

CPC current components for phase *a* are shown in Fig. 3.26.

Simulation results for transient state in case of sinusoidal supply with zero internal impedance and symmetrical load are shown in Fig. 3.27.

CPC current components for phase *a* (Fig. 3.28).

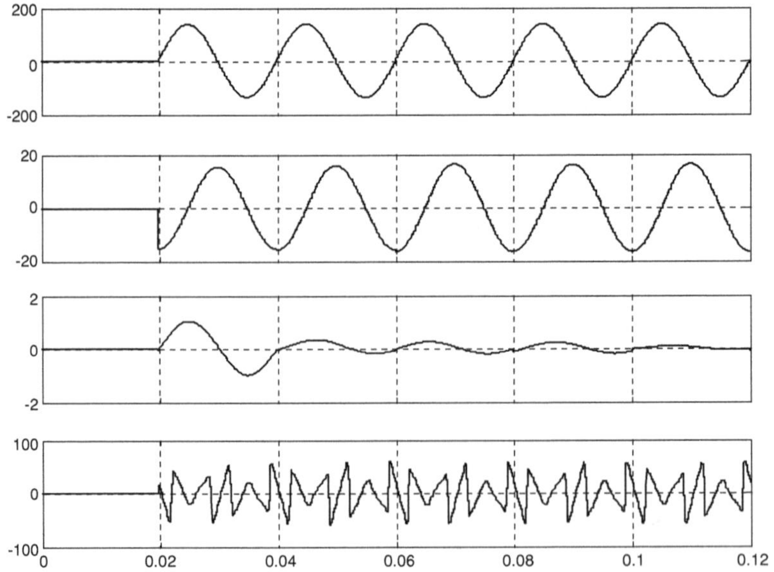

Fig. 3.34 CPC components, i.e. active, reactive, unbalance current and current generated for phase *a*

Simulation results for steady state in case of sinusoidal supply with zero internal impedance and non-symmetrical load are shown in Figs. 3.27, 3.29.

CPC current components for phase *a* (Fig. 3.30).

Simulation results for transient state in case of sinusoidal supply with zero internal impedance and non-symmetrical load are shown in Fig. 3.31.

CPC current components for phase *a* (Fig. 3.32).

Simulation results for steady state in case of non-sinusoidal supply with non-zero internal impedance and symmetrical load are shown in Fig. 3.33.

CPC current components for phase *a* (Fig. 3.34).

Simulation results for transient state in case of non-sinusoidal supply with non-zero internal impedance and symmetrical load are shown in Fig. 3.35.

CPC current components, phase *a* (Fig. 3.36).

Simulation results for steady state in case of non-sinusoidal supply with non-zero internal impedance and non-symmetrical load are shown in Fig. 3.37.

CPC current components for phase *a* (Fig. 3.38).

Simulation results for transient state in case of non-sinusoidal supply with non-zero internal impedance and non-symmetrical load are shown in Fig. 3.39.

CPC current components for phase *a* (Fig. 3.40).

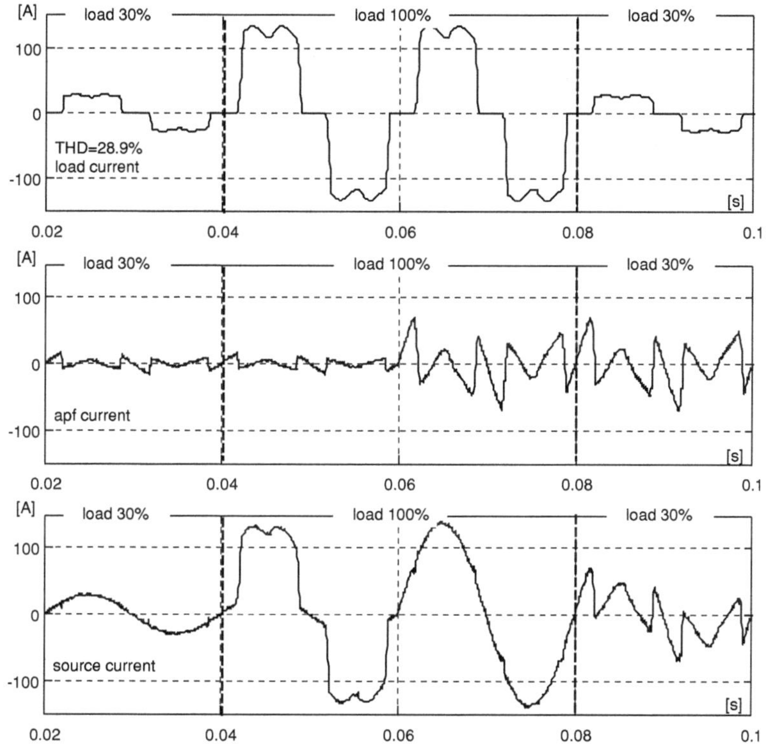

Fig. 3.35 Simulation results, phase *a*

3.1.3.3 Conclusions

- Properties of discussed control algorithm are similar to those of algorithm using Budeanu theory. This comes of using frequency domain in system analysis.
- According to its author, CPC theory is currently the only theory which makes it possible to relate current components used to actual physical phenomena occurring in circuits with periodical and non-sinusoidal waveforms. However, the orthogonal distribution used by this theory is debatable in circuits with real (non-ideal) voltage source (i.e. with sources characterised by non-zero internal impedance).
- Hardware implementation of this algorithm is very complicated (even if DSPs are used) and therefore it is very rarely applied in active compensation systems (APFs).

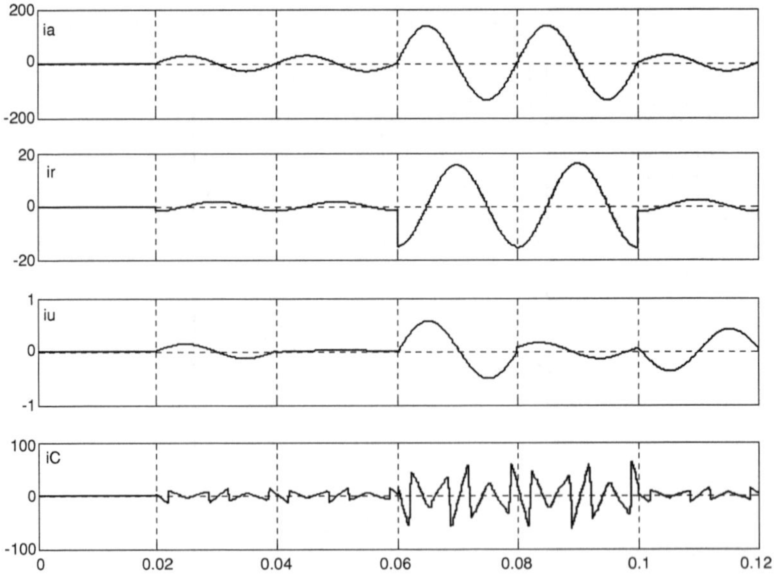

Fig. 3.36 CPC components, i.e. active, reactive, unbalance current and current generated for phase *a*

3.1.4 Control Strategy Based on pq Theory

This algorithm uses time domain to determine standard compensating currents. *pq* theory is not universally acknowledged as power theory, since it has been originally elaborated for three-phase circuits only. However, it is widely applied as a tool to eliminate undesired (adverse) components of supply currents. Block diagram of the algorithm is shown in Fig. 3.41.

The acquired instantaneous values of phase currents and voltages (block {1}) are transformed from natural three-phase coordinate system to Cartesian coordinate system (block {2}) in accordance with Eq. 3.12. A simplifying assumption has been that energy transfer is achieved with three-wire system, this allows for elimination of zero-sequence component.

$$\begin{bmatrix} f_\alpha[n] \\ f_\beta[n] \end{bmatrix} = \sqrt{\frac{2}{3}} \begin{bmatrix} 1 & -\frac{1}{2} & -\frac{1}{2} \\ 0 & \frac{\sqrt{3}}{2} & -\frac{\sqrt{3}}{2} \end{bmatrix} \begin{bmatrix} f_\mathrm{a}[n] \\ f_\mathrm{b}[n] \\ f_\mathrm{c}[n] \end{bmatrix} \tag{3.23}$$

Next, power state of the system is identified (block {3}) and this calls for determination of instantaneous values of powers present in the system, i.e.:

$$\begin{bmatrix} p[n] \\ q[n] \end{bmatrix} = \begin{bmatrix} v_\alpha[n] & v_\beta[n] \\ -v_\beta[n] & v_\alpha[n] \end{bmatrix} \begin{bmatrix} i_\alpha[n] \\ i_\beta[n] \end{bmatrix} \tag{3.24}$$

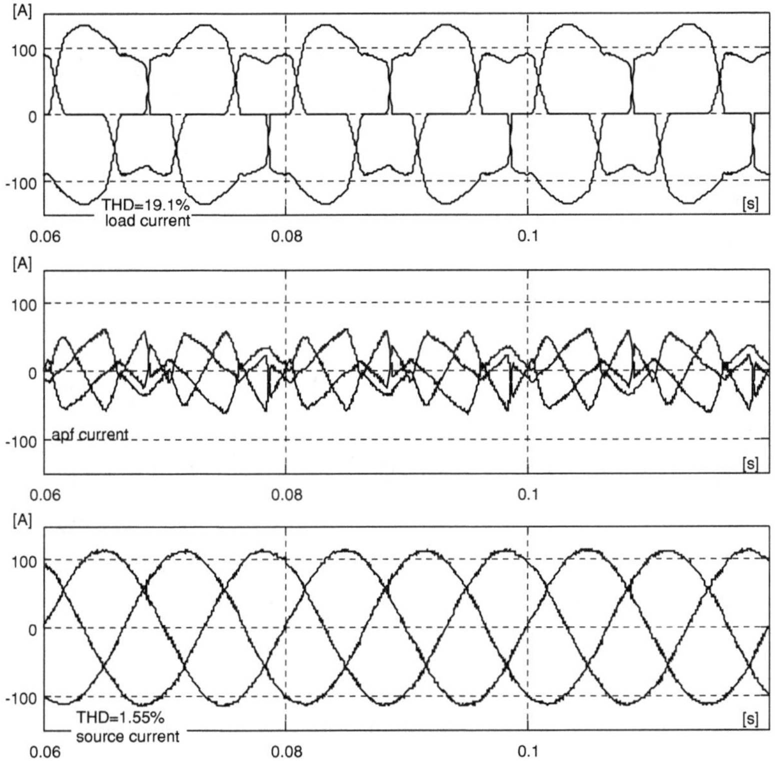

Fig. 3.37 Simulation results, THD coefficients relate to phase *a*

Afterwards, decomposition of instantaneous powers into components must be done. The individual components are related to positive-sequence fundamental harmonic, to currents' higher harmonics and to negative-sequence harmonic [22, 23]:

$$\begin{bmatrix} p[n] \\ q[n] \end{bmatrix} = \begin{bmatrix} \bar{p} + \tilde{p}[n] \\ \bar{q} + \tilde{q}[n] \end{bmatrix} \tag{3.25}$$

where:

$\bar{p}$—constant component of instantaneous active power related to positive-sequence fundamental harmonic of source current. This component corresponds to instantaneous value of conventionally defined active current [21], i.e. current with minimum RMS-value ensuring required active power of the load (this is in accordance with Fryze's definition);

$\tilde{p}[n]$—variable component of instantaneous active power related to negative-sequence harmonic (with 2ω pulsation) [20] and higher harmonics of source

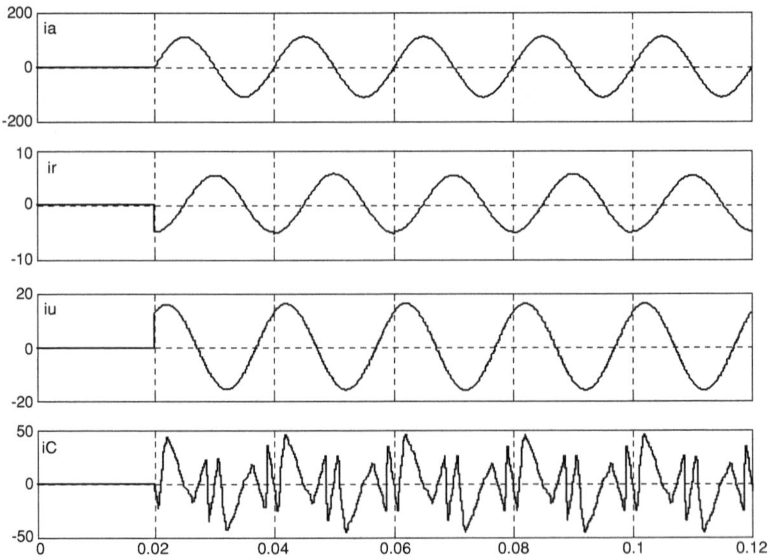

Fig. 3.38 CPC components, i.e. active, reactive, unbalance current and current generated for phase *a*

current; it corresponds to transfer of energy (per time unit) between source and load;

$\bar{q}$—constant component of imaginary power related to positive-sequence fundamental harmonic of source current. In case of three-phase symmetrical sinusoidal system (voltages and currents both), value of constant component of imaginary power is equivalent to conventionally defined reactive power based on average value;

$\tilde{q}[n]$—variable component of instantaneous imaginary power related to negative-sequence harmonic (with 2 pulsation) and higher harmonics of source current.

Decomposition of the components is conducted using digital filters with infinite pulse response (block {4}).The principal advantage of Cartesian co-ordinates α-β is the possibility of simple notation of compensating currents resultant equations. Before active filter is switched on, the source current transformed to α-β system may be expressed as:

$$\begin{bmatrix} i_\alpha[n] \\ i_\beta[n] \end{bmatrix} = \frac{1}{v_\alpha^2[n] + v_\beta^2[n]} \begin{bmatrix} v_\alpha[n] & -v_\beta[n] \\ v_\beta[n] & v_\alpha[n] \end{bmatrix} \begin{bmatrix} \overline{p} + \tilde{p}[n] \\ \overline{q} + \tilde{q}[n] \end{bmatrix} \tag{3.26}$$

For symmetrical voltage supply source:

$$v_\alpha[n]^2 + v_\beta^2[n]^2 = 3|V_a|^2 = \text{const} \tag{3.27}$$

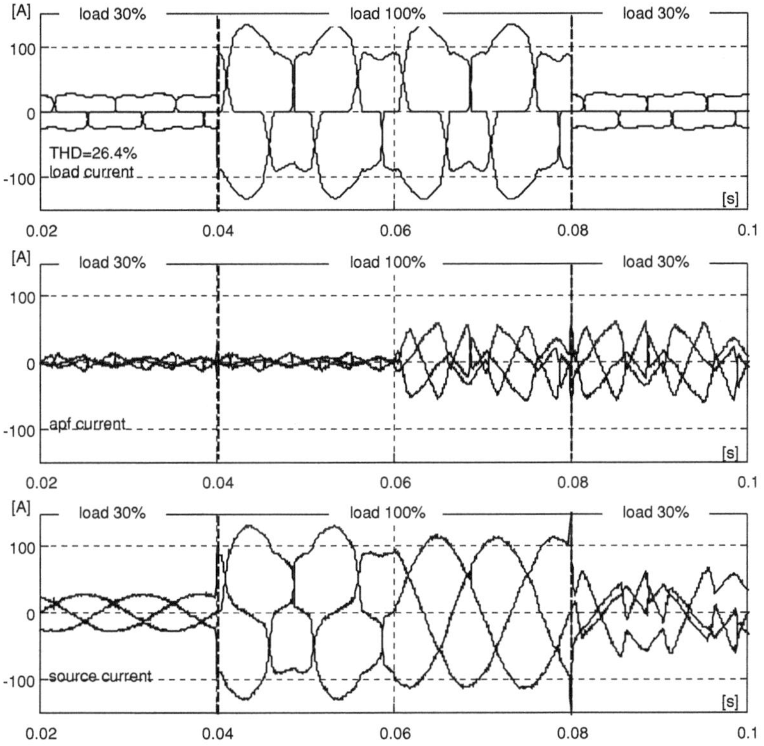

Fig. 3.39 Simulation results, THD coefficient relates to phase *a*

Reference compensation currents (block {5}) may be determined with the help of formula:

$$\begin{bmatrix} i_\alpha[n] \\ i_\beta[n] \end{bmatrix} = \frac{1}{v_\alpha[n]^2 + v_b[n]^2} \begin{bmatrix} v_\alpha[n] & -v_\beta[n] \\ v_\beta[n] & v_\alpha[n] \end{bmatrix} \begin{bmatrix} p_k[n] \\ q_k[n] \end{bmatrix} \tag{3.28}$$

Depending on which component should be eliminated, p_k and q_k must be substituted with quantities taken from Table 3.1.

Compensator currents obtained in this way must next be transformed from Cartesian **α-β** co-ordinate system to natural three-phase circuit ***a*-*b*-*c*** (block {6}). Next, using *PI* controller (block {7}) and hardware PWM circuits (block {8}) inverter's power electronic valves will be controlled.

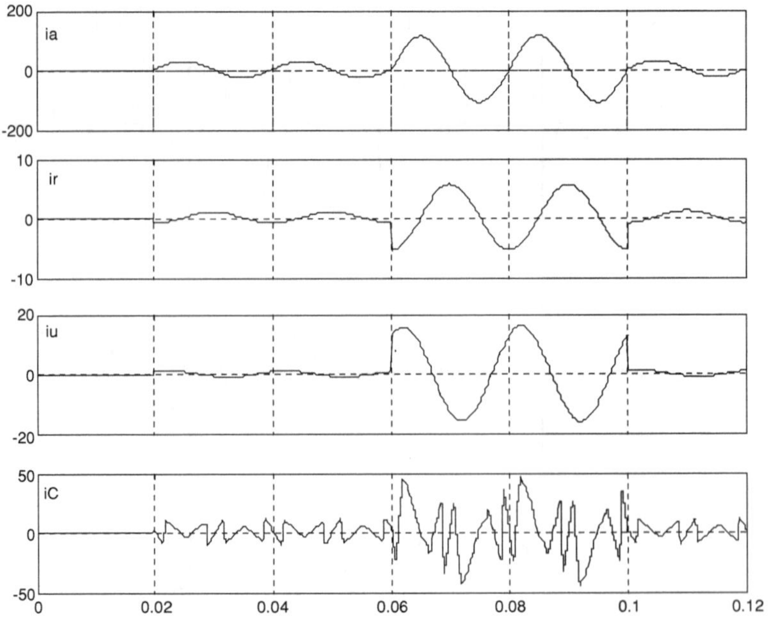

Fig. 3.40 CPC components, i.e. active, reactive, unbalance current and current generated for phase *a*

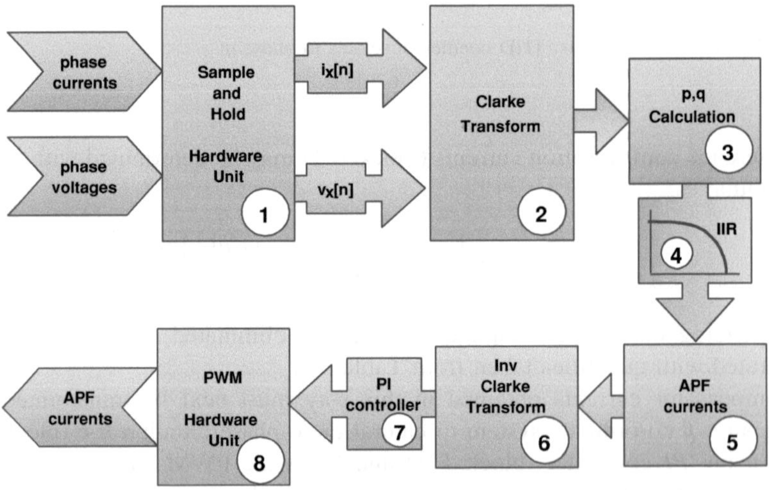

Fig. 3.41 Block diagram of control algorithm using *pq* theory

Table 3.1 Source current components

Eliminated source current component	$p_k[n]$	$q_k[n]$
Related to instantaneous imaginary power	0	$q[n]$
Negative-sequence component and higher harmonics	$\tilde{p}[n]$	$\tilde{q}[n]$
Related to constant component of instantaneous imaginary power	0	$\bar{q}$
Related to instantaneous imaginary power and higher harmonics	$\tilde{p}$	q
Negative-sequence component	$p_{2\omega}[n]$	$q_{2\omega}[n]$
Related to higher harmonics components	$p_h[n]$	$q_h[n]$
Related to variable component of instantaneous active power	$\tilde{p}[n]$	0

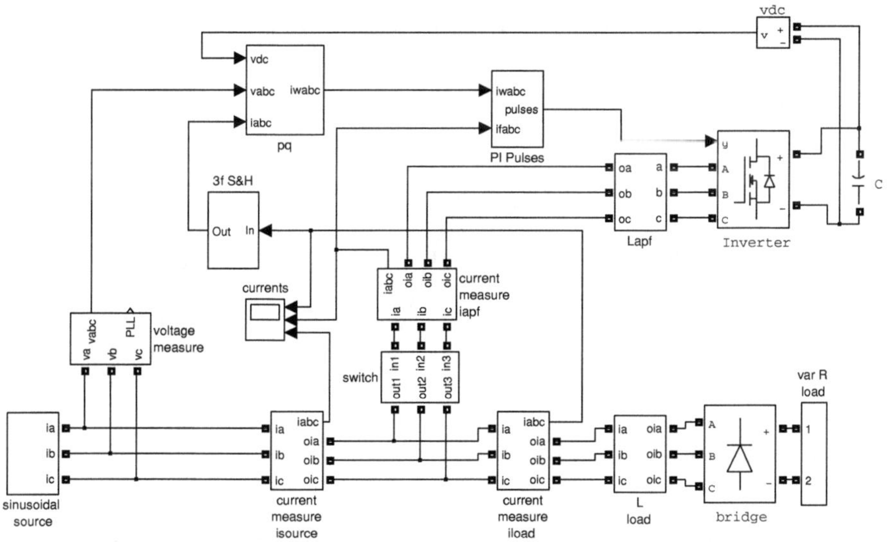

Fig. 3.42 Block diagram of model elaborated in Matlab/Simulink environment

3.1.4.1 Model of Source-APF-Load Circuit in Matlab/Simulink Environment

Model elaborated for analysis of control algorithm utilising *pq* theory is shown in Fig. 3.42.

3.1.4.2 Simulation Results

Circuit shown in Fig. 3.42 has been analysed under different operating conditions (same as in case of algorithms described previously). Simulation results for steady state in case of sinusoidal source with zero internal impedance and symmetrical load are shown in Fig. 3.43.

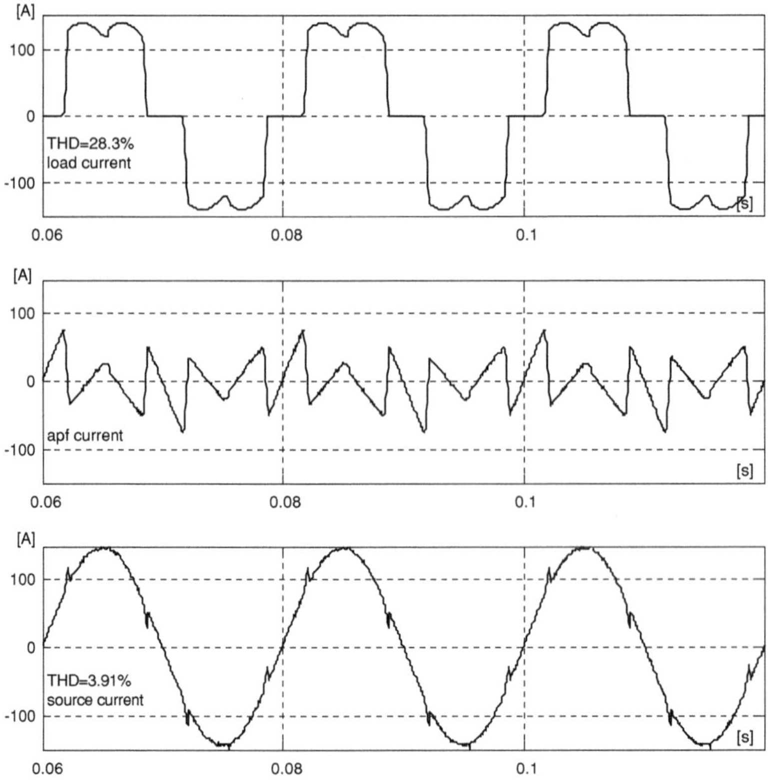

Fig. 3.43 Simulation results, phase *a*

Simulation results for transient state in case of sinusoidal source with zero internal impedance and symmetrical load are shown in Fig. 3.44.

Simulation results for steady state in case of sinusoidal source with zero internal impedance and non-symmetrical load are shown in Fig. 3.45.

Simulation results for transient state in case of sinusoidal source with zero internal impedance and non-symmetrical load are shown in Fig. 3.46.

Simulation results for steady state in case of non-sinusoidal source with non-zero internal impedance and symmetrical load are shown in Fig. 3.47. In this case source current components defined in accordance with Fryze's theory are no longer orthogonal.

Simulation results for transient state in case of non-sinusoidal source with non-zero internal impedance and symmetrical load are shown in Fig. 3.48.

Simulation results for steady state in case of non-sinusoidal source with non-zero internal impedance and non-symmetrical load are shown in Fig. 3.49.

Simulation results for transient state in case of non-sinusoidal source with non-zero internal impedance and non-symmetrical load are shown in Fig. 3.50.

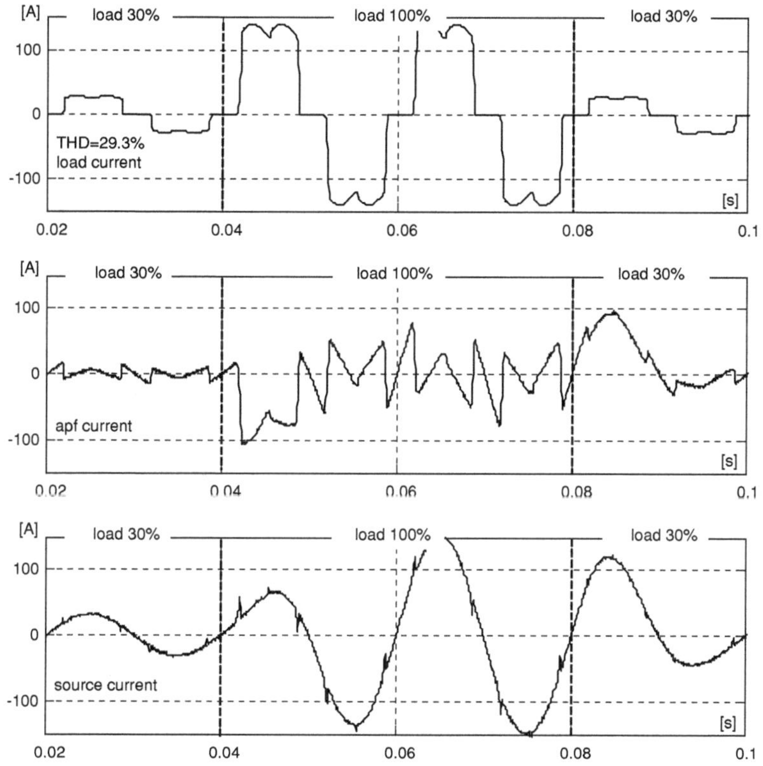

Fig. 3.44 Simulation results, phase *a*

3.1.4.3 Conclusions

- *pq* Theory makes it possible to design effective algorithms for active filter (active compensator) control
- The dynamics of control system is strongly associated with type of digital filter used. If fast response filter is applied, then load variations may be closely followed; however, some variations (fluctuations) in instantaneous power may still arise in the circuit. That is why filter should be dedicated to specific circuit
- In accordance with the requirements the system may operate as: active filter of higher harmonics, compensator symmetrising generalised load seen from source terminals or active compensator eliminating all undesirable components of instantaneous power
- If we aim at elimination of components related to instantaneous imaginary power only, the compensator does not have to be equipped with energy storage elements (in ideal case $pk[n] = 0$, which means that there is no exchange of active power between compensator and load), but uses simply switching elements

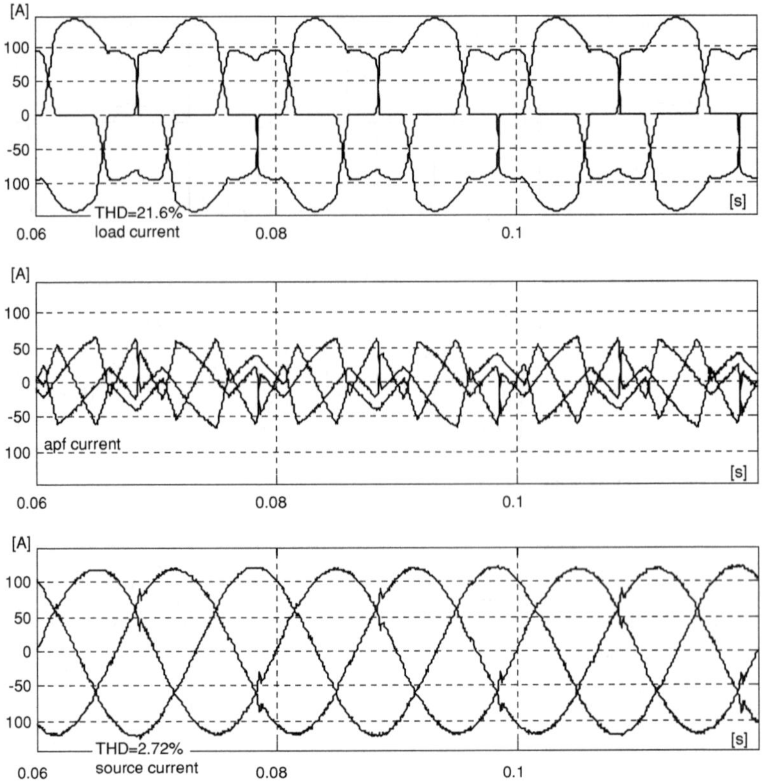

Fig. 3.45 Simulation results, THD coefficients relate to phase *a*

- Drawbacks and limits of this theory must be noted; among the most important is the fact that it may be applied in three-phase systems only. Besides, it is difficult to calculate constant components of instantaneous active and imaginary powers fast enough and incorrect standard compensating currents are generated in case of periodical and deformed or non-symmetrical supply signal.

3.1.5 Control Strategy Based on Extension pq Theory

Simple and effective control method based on instantaneous power theory "*pq*" and proposed by Akagi [24] does not work properly in systems with non-symmetrical supply voltage. Since systems with slight supply asymmetries occur quite often, then some other approach to working out active filter control strategy is necessary. In 1995, Komatsu and Kawabata [25–27] presented instantaneous power theory called "extension *pq*". They propose a more general compensating approach, ensuring correct operation of control system even if asymmetries are present.

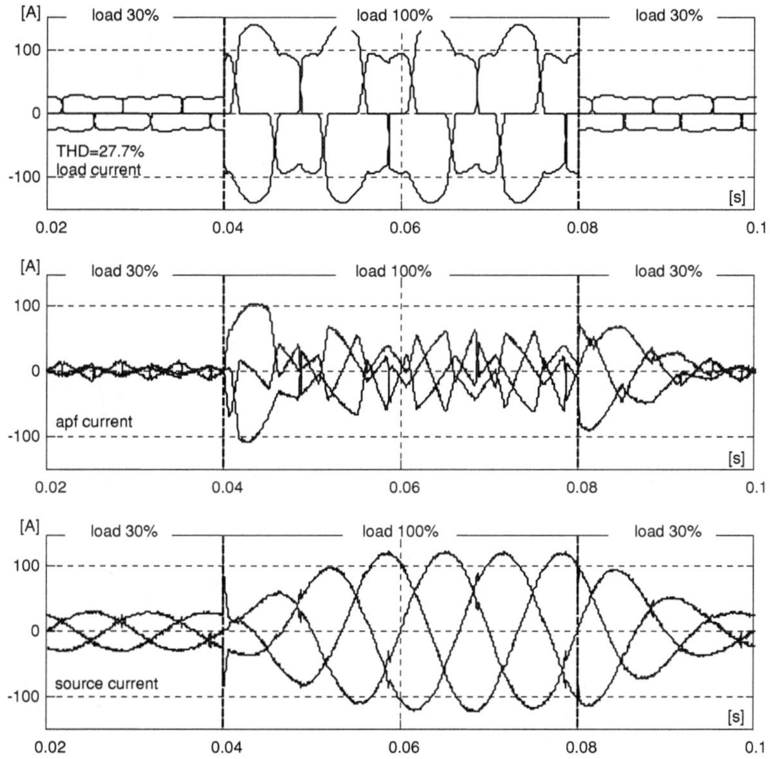

Fig. 3.46 Simulation results, THD coefficients relate to phase *a*

Block diagram of this algorithm is shown in Fig. 3.51.

Phase current and voltage waveforms are subjected to sampling process (block {1}). The acquired instantaneous voltage values are then delayed (block {2}) by one-fourth of period; as a result, the so-called transverse (perpendicular) voltages are obtained. These values and samples of phase voltages and currents instantaneous values are the basis for calculation of instantaneous power values (block {3}).

$$\begin{aligned} p[n] &= v_{\mathrm{a}}[n]i_{\mathrm{a}}[n] + v_{\mathrm{b}}[n]i_{\mathrm{b}}[n] + v_{\mathrm{c}}[n]i_{\mathrm{c}}[n] \\ q[n] &= v'_{\mathrm{a}}[n]i_{\mathrm{a}}[n] + v'_{\mathrm{b}}[n]i_{\mathrm{b}}[n] + v'_{\mathrm{c}}[n]i_{\mathrm{c}}[n] \end{aligned} \tag{3.29}$$

where: $v'_{\mathrm{a,b,c}}[n]$—transverse (perpendicular) voltages.

Additionally, in case of common three-phase three-wire systems we may apply Kirchoff's first law and create the following equation:

$$i_a[n] + i_b[n] + i_c[n] = 0 \tag{3.30}$$

thus reducing instantaneous power equation to the form:

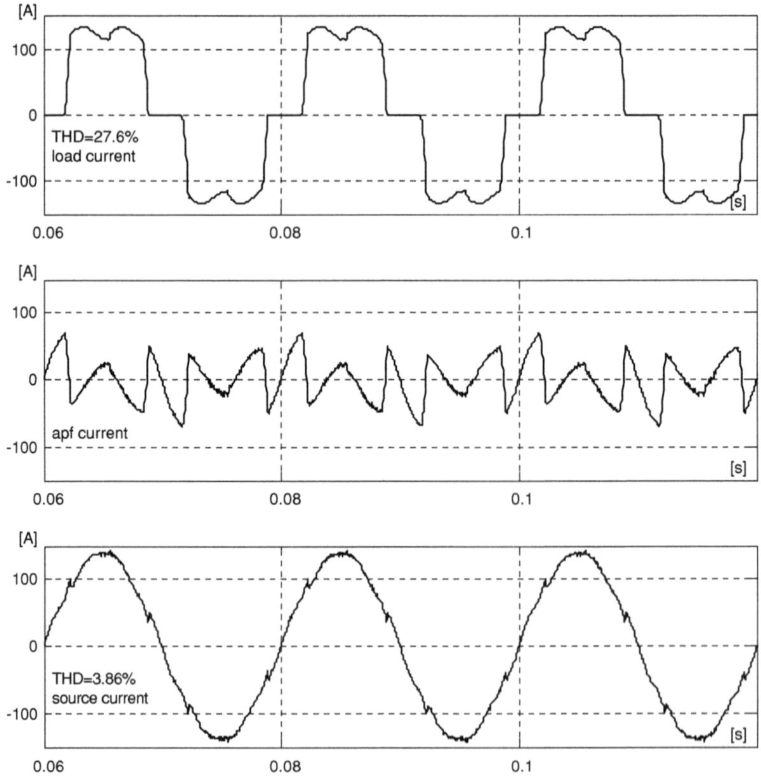

Fig. 3.47 Simulation results, phase *a*

$$\begin{bmatrix} p[n] \\ q[n] \end{bmatrix} = \begin{bmatrix} v_a[n] - v_c[n] & v_b[n] - v_c[n] \\ v'_a[n] - v'_c[n] & v'_b[n] - v'_c[n] \end{bmatrix} \begin{bmatrix} i_a[n] \\ i_b[n] \end{bmatrix} \tag{3.31}$$

Digital low-pass filter (block {4}) eliminates variable component of instantaneous active power. By using power components, instantaneous values of compensating currents are then calculated (block {5}).

$$\begin{aligned} \begin{bmatrix} i_{fa}[n] \\ i_{fb}[n] \end{bmatrix} &= \begin{bmatrix} i_a[n] \\ i_b[n] \end{bmatrix} - \frac{1}{\Delta} \begin{bmatrix} v'_b[n] - v'_c[n] & v_c[n] - v_b[n] \\ v'_c[n] - v'_a[n] & v_a[n] - v_c[n] \end{bmatrix} \begin{bmatrix} \overline{p} \\ 0 \end{bmatrix} \\ &= \begin{bmatrix} i_a[n] \\ i_b[n] \end{bmatrix} - \frac{\overline{p}}{\Delta} \begin{bmatrix} v'_b[n] - v'_c[n] \\ v'_c[n] - v'_a[n] \end{bmatrix}; \; i_{fc}[n] = -i_{fa}[n] - i_{fb}[n] \end{aligned} \tag{3.32}$$

where:

$$\Delta = (v_a[n] - v_c[n])(v'_b[n] - v'_c[n]) - (v'_a[n] - v'_c[n])(v_b[n] - v_c[n]) \tag{3.33}$$

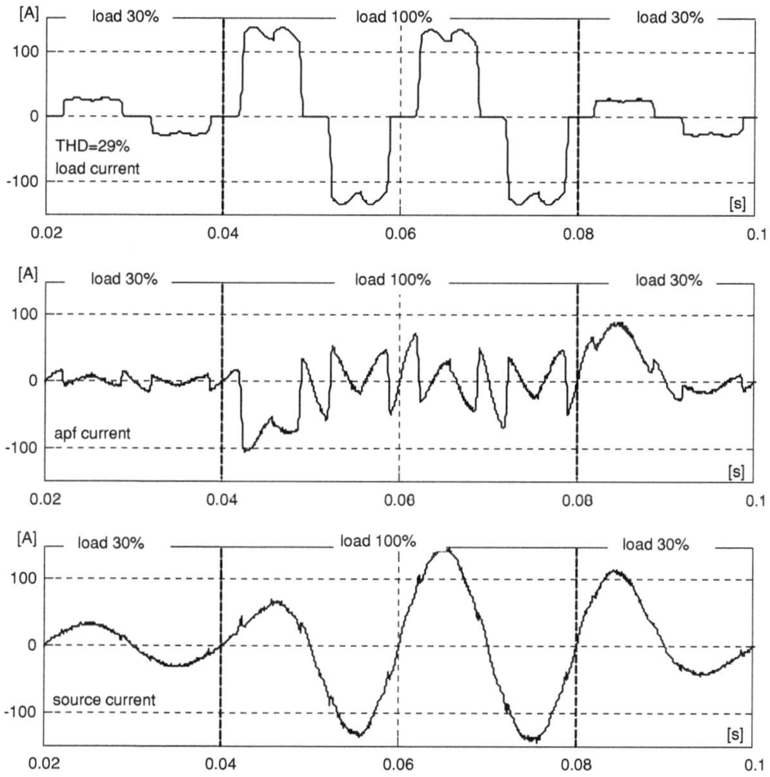

Fig. 3.48 Simulation results, phase *a*

Standards determined in accordance with this procedure are then input to source-APF-load system via PI controller block ({6}) and hardware PWM module (block {7}) controlling voltage inverter.

3.1.5.1 Source-APF-Load Model in Matlab/Simulink Environment

The elaborated model is shown in Fig. 3.52.

3.1.5.2 Simulation Results

Circuit shown in Fig. 3.52 has been analysed under different operating conditions (same as in case of algorithms described before). Simulation results for steady state in case of sinusoidal source with zero internal impedance and symmetrical load are shown in Fig. 3.53.

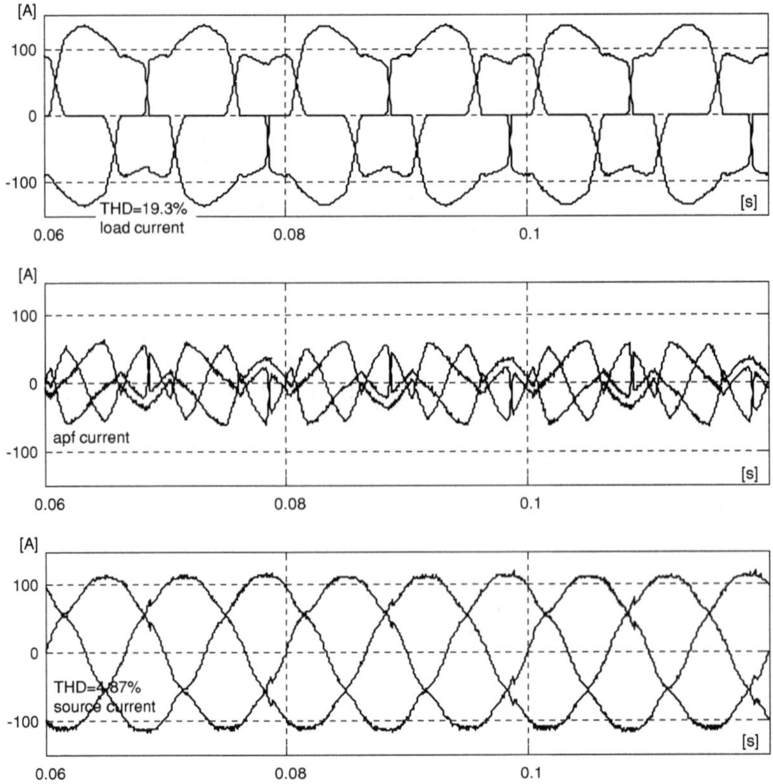

Fig. 3.49 Simulation results, THD coefficients relate to phase *a*

Simulation results for transient state in case of sinusoidal source with zero internal impedance and symmetrical load are shown in Fig. 3.54.

Simulation results for steady state in case of sinusoidal source with zero internal impedance and non-symmetrical load are shown in Fig. 3.55.

Simulation results for transient state in case of sinusoidal source with zero internal impedance and non-symmetrical load are shown in Fig. 3.56.

Simulation results for steady state in case of non-sinusoidal source with non-zero internal impedance and symmetrical load are shown in Fig. 3.57. In this case source current components defined in accordance with Fryze's theory are no longer orthogonal.

Simulation results for transient state in case of non-sinusoidal source with non-zero internal impedance and symmetrical load are shown in Fig. 3.58.

Simulation results for steady state in case of non-sinusoidal source with non-zero internal impedance and non-symmetrical load are shown in Fig. 3.59.

Simulation results for transient state in case of non-sinusoidal source with non-zero internal impedance and non-symmetrical load are shown in Fig. 3.60.

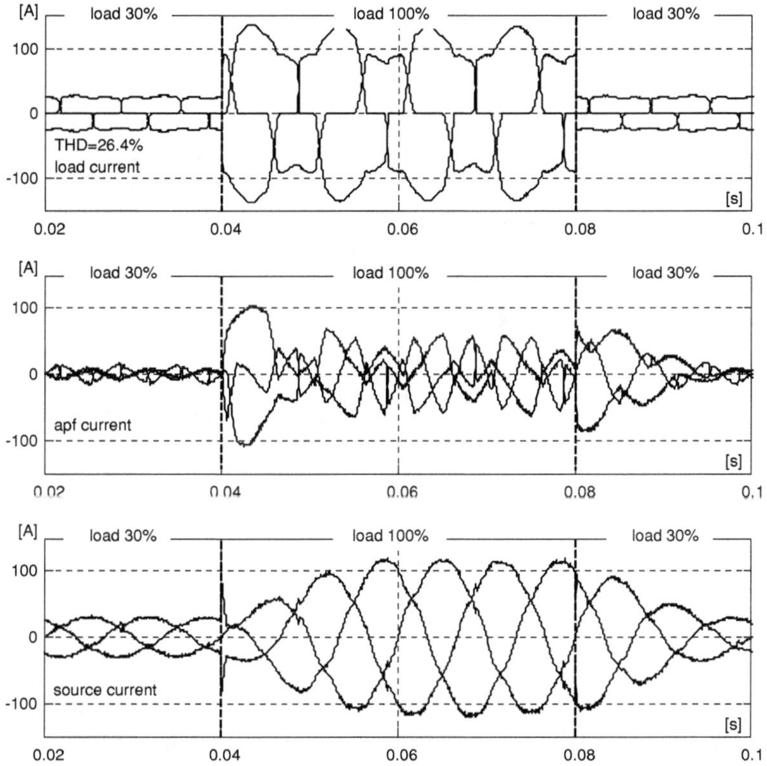

Fig. 3.50 Simulation results, THD coefficients relate to phase *a*

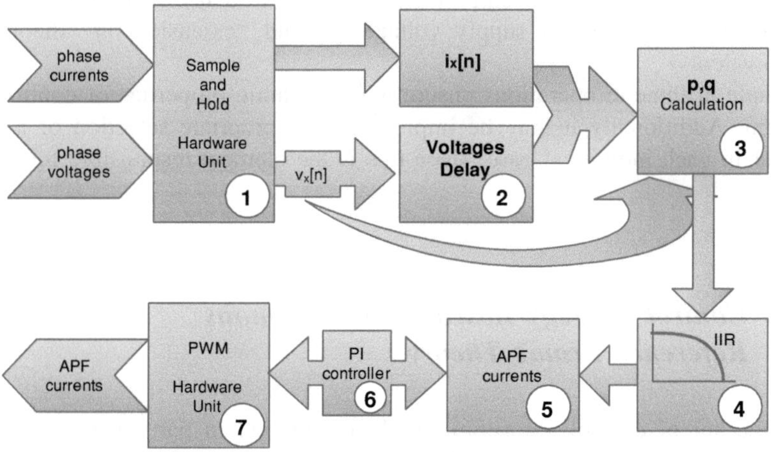

Fig. 3.51 Block diagram of control algorithm using "extension pq" theory

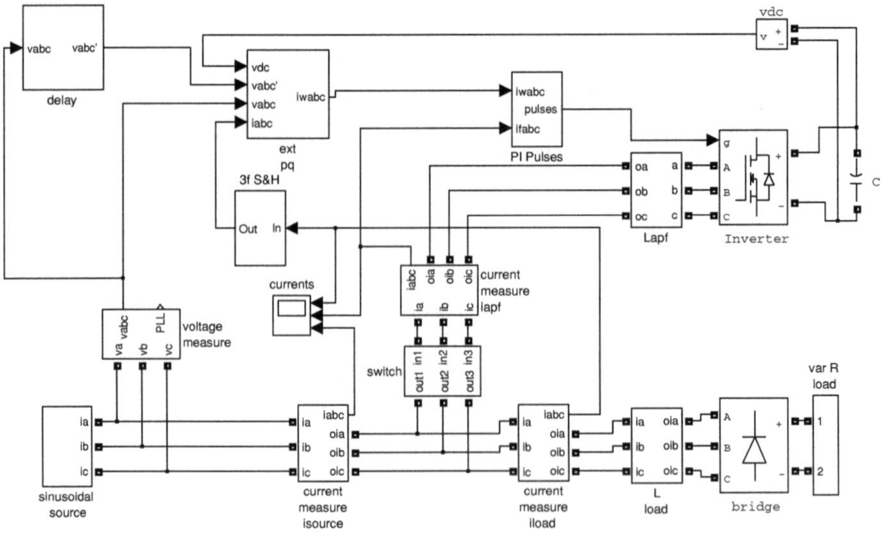

Fig. 3.52 Block diagram of the model in Matlab/Simulink environment

3.1.5.3 Conclusions

- Contrary to active filter control algorithm using instantaneous power theory *pq* [20, 24, 28], algorithm based on instantaneous power theory "extension *pq*" [25, 26] eliminated deformations of current sources even in case of non-symmetrical supply
- If voltage is deformed, then generated standard currents may be incorrect. This is due to applied delay circuit which is adapted to delay of the first harmonic
- In case of symmetrical supply voltage, *pq* and "extension *pq*" theories are equivalent
- Simple arithmetic operations ensure good dynamic properties of control algorithm. Additionally it may be improved by appropriate selection of a digital filter in each individual case, since this is the element responsible for greatest delays.

3.1.6 Control Strategy Based on Synchronous Reference Frame Theory

This variant of control algorithm should be applied in particular when supply voltage is deformed or badly suffused with interference. Block diagram of the algorithm is shown in Fig. 3.61.

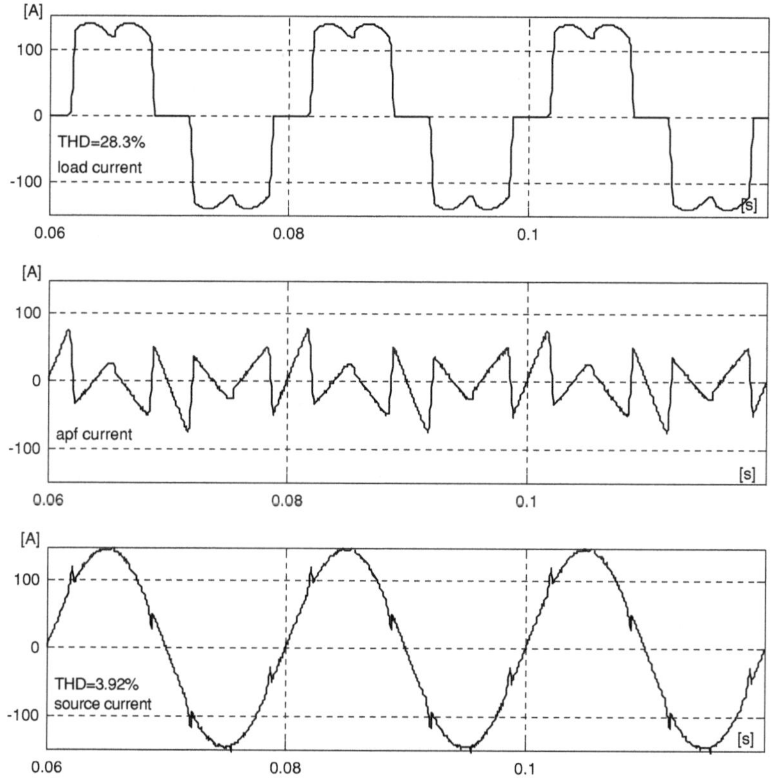

Fig. 3.53 Simulation results, phase *a*

This approach uses transformation [29] of input signal vectors from natural three-phase system (***a***-***b***-***c***) into rotating coordinates system ***d***-***q***. Sampled phase currents waveforms (block {1}) are initially subjected to transformation into stationary ***α***-***β*** system (block {2}), same as in *pq* algorithm, and afterwards to ***d***-***q*** system rotating at speed ω (block {3}). Values in the new coordinate system will be obtained from the equation:

$$\begin{bmatrix} f_d[n] \\ f_q[n] \end{bmatrix} = \begin{bmatrix} \cos\theta[n] & \sin\theta[n] \\ -\sin\theta[n] & \cos\theta[n] \end{bmatrix} \begin{bmatrix} f_\alpha[n] \\ f_\beta[n] \end{bmatrix}. \tag{3.34}$$

This transformation is also known as Park transformation [30]. The advantage of this transform lies in the fact that the system does not require identification of energy state by determination of instantaneous values of active and reactive powers. If course of $\cos\theta[n]$ function is the same as that of phase *a* voltage fundamental harmonic, then ***d***-***q*** system rotates synchronously with supply voltage fundamental harmonic. The synchronisation is commonly implemented by using phase loop (PLL) synchronised with voltage fundamental harmonic waveform.

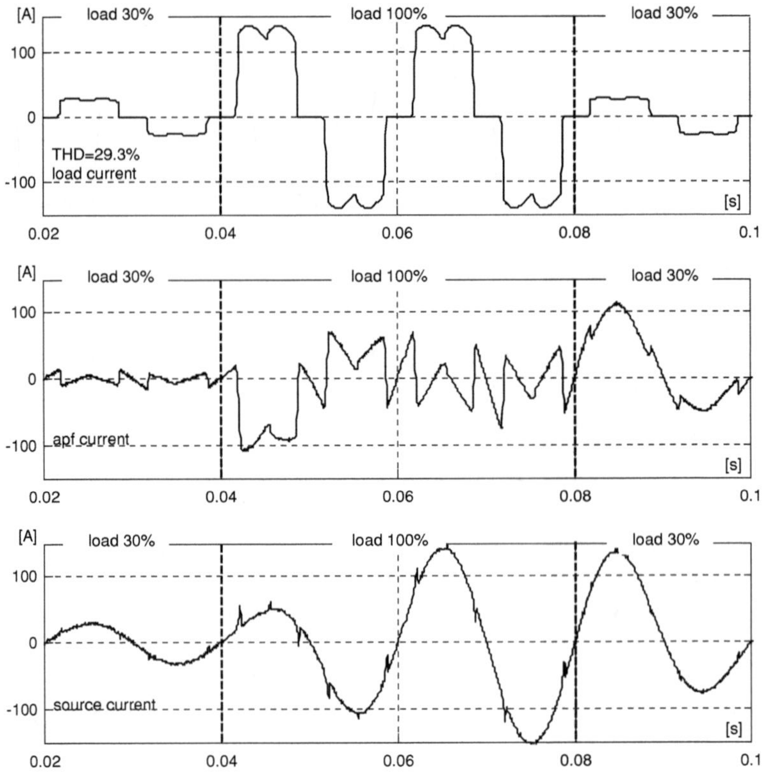

Fig. 3.54 Simulation results, phase a

In this reference system the quantities being in phase with supply voltage fundamental harmonic (i.e. active component of source current) are represented by time-invariant values. Thus average value of $\boldsymbol{d}$-axis current component corresponds to source current active component in rotating coordinates system [31]. That is why the active filter current waveforms may be calculated directly using signal filters (block {4}), provided phase currents in new co-ordinate system are known. These signals must be subsequently subjected to inverse Park and Clarke transformations (blocks {5} and {6}) and, a result, we obtain:

$$\begin{bmatrix} i_{fa}[n] \\ i_{fb}[n] \\ i_{fc}[n] \end{bmatrix} = \sqrt{\frac{2}{3}} \begin{bmatrix} 1 & 0 \\ -\frac{1}{2} & \frac{\sqrt{3}}{2} \\ -\frac{1}{2} & -\frac{\sqrt{3}}{2} \end{bmatrix} \begin{bmatrix} \cos\theta[n] & -\sin\theta[n] \\ \sin\theta[n] & \cos\theta[n] \end{bmatrix} \begin{bmatrix} \tilde{i}_{\mathrm{d}}[n] \\ i_{\mathrm{q}}[n] \end{bmatrix} \tag{3.35}$$

Afterwards the phase currents of APF are input into the circuit via PI controller (block {7}) and hardware PWM module (block {8}.

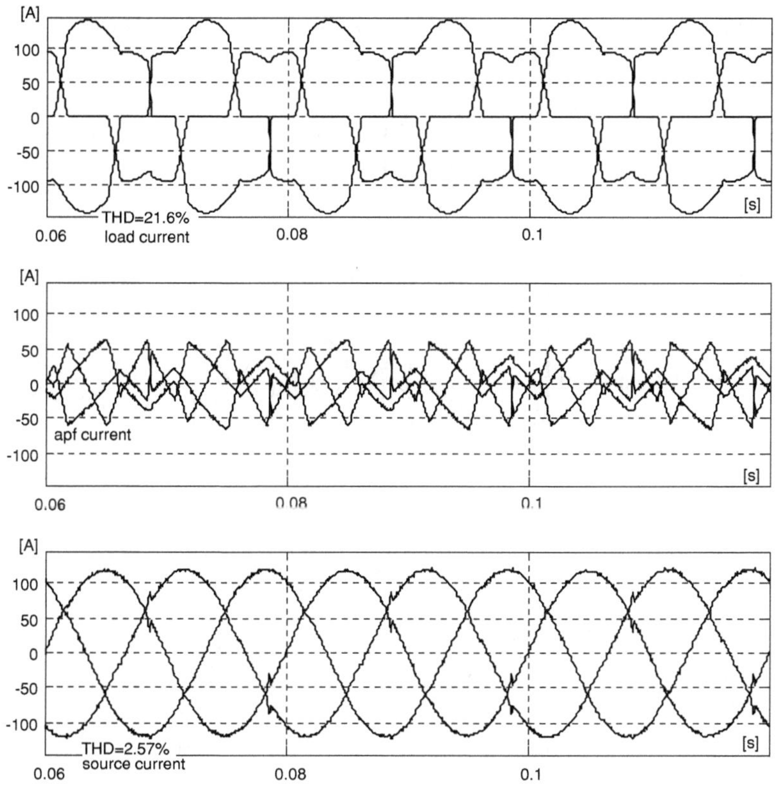

Fig. 3.55 Simulation results, THD coefficients relate to phase *a*

3.1.6.1 Model of Source-APF-Load System in Matlab/Simulink Environment

The elaborated model is shown in Fig. 3.62.

3.1.6.2 Simulation Results

Circuit shown in Fig. 3.62 has been analysed under different operating conditions (same as in case of algorithms described previously). Simulation results for steady state in case of sinusoidal source with zero internal impedance and symmetrical load are shown in Fig. 3.63.

Simulation results for transient state in case of sinusoidal source with zero internal impedance and symmetrical load are shown in Fig. 3.64.

Simulation results for steady state in case of sinusoidal source with zero internal impedance and non-symmetrical load are shown in Fig. 3.65.

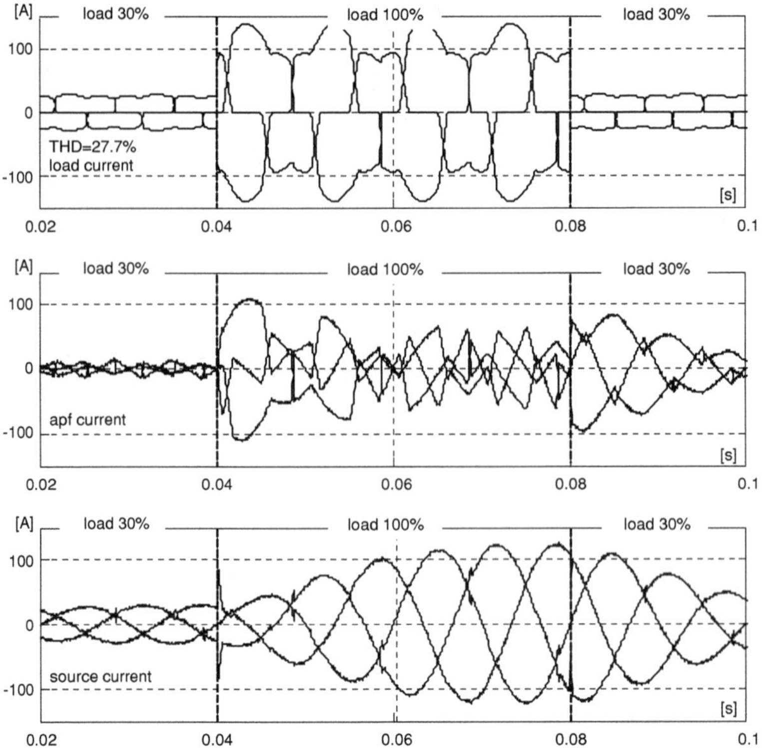

Fig. 3.56 Simulation results, THD coefficients relate to phase *a*

Simulation results for transient state in case of sinusoidal source with zero internal impedance and non-symmetrical load are shown in Fig. 3.66.

Simulation results for steady state in case of non-sinusoidal source with non-zero internal impedance and symmetrical load are shown in Fig. 3.67. In this case source current components defined in accordance with Fryze's theory are no longer orthogonal.

Simulation results for transient state in case of non-sinusoidal source with non-zero internal impedance and symmetrical load are shown in Fig. 3.68.

Simulation results for steady state in case of non-sinusoidal source with non-zero internal impedance and non-symmetrical load are shown in Fig. 3.69.

Simulation results for transient state in case of non-sinusoidal source with non-zero internal impedance and non-symmetrical load are shown in Fig. 3.70.

3.1.6.3 Conclusions

- Eliminating the necessity of determining active and reactive power instantaneous values causes significant reduction in the number of required mathematical operations in each cycle where compensator currents are computed.

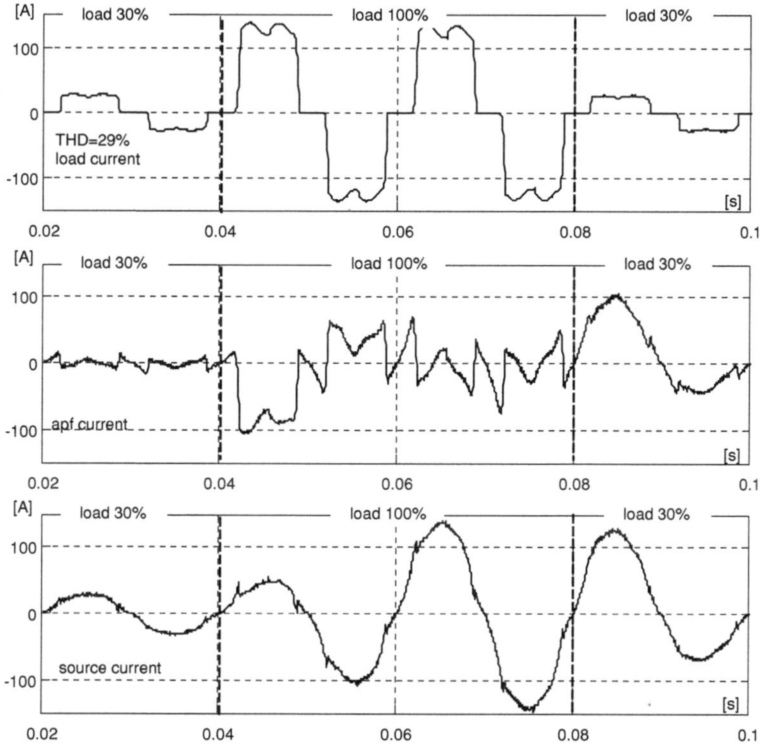

Fig. 3.57 Simulation results, phase *a*

Reduced number of operations results in decreased processing time, so that the dynamics of the control system is improved

- Using phase loop (PLL) in the system ensures correct determination of compensating currents even if periodic supply voltage is deformed or full of interference
- Simple structure of the control system allows for easy hardware implementation using DSPs
- In case of sinusoidal and symmetrical supply this method is equivalent [29] to method proposed by Akagi [24].

3.2 V_{dc} Voltage Controller

Each of the algorithms presented so far requires additional constant voltage (provided by capacitor, i.e. DC voltage link within the voltage inverter). This issue has not been discussed so far, since it does not depend on choice of algorithm.

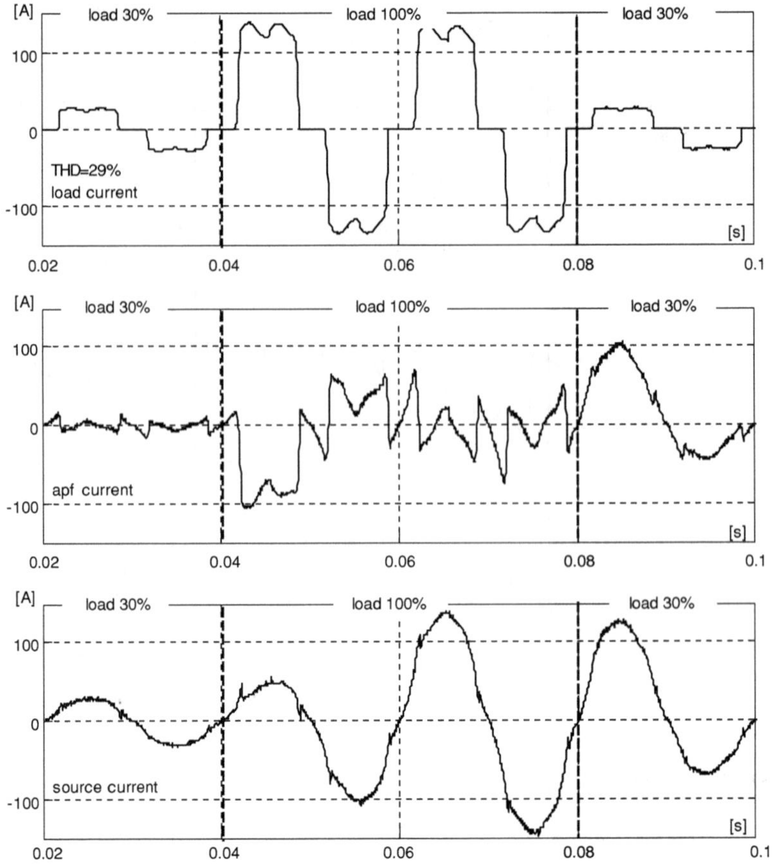

Fig. 3.58 Simulation results, phase *a*

Signals of standard compensating currents determined as a result of any algorithm, must be corrected in such a way that voltage inverter charging DC link capacitor should keep its voltage at a constant level. In transient states (e.g. when load changes), great difference between load power and standard active power (or active current—depends on control algorithm adopted by the control circuit) exists. This difference is caused by finite time needed to determine correct standard [32]. This results in capacitor voltage variations and, in its turn, generates the need for appropriate V_{dc} control. V_{dc} controller introduces corrections to standard compensating currents on the basis of measured existing voltage and set reference voltage. The selection of V_{cd} voltage controller parameters depends on load power variations level. The controllers used should not react too quickly to small changes of capacitor voltage, since this might lead to unwanted distortion of supply currents. On the other hand, in steady-state conditions keeping capacitor voltage constant requires supplying the inverter with constant active power compensating

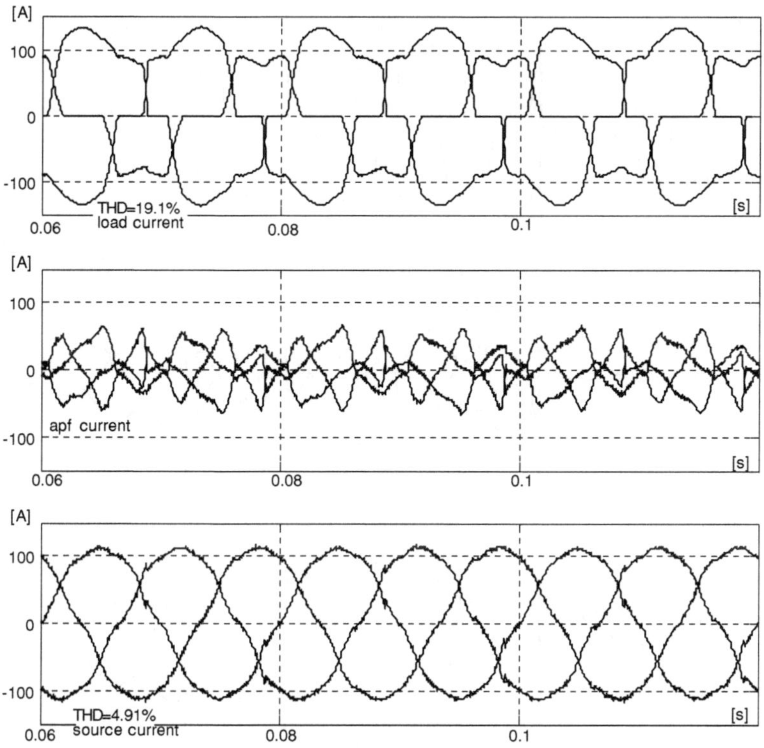

Fig. 3.59 Simulation results, THD coefficients relate to phase *a*

power losses in the inverter and active power due to generated current waveform. Usually standard *PI* controllers are used for V_{dc} control, less often other designs are applied such as fuzzy logic controllers.

3.3 Experimental Results of DSP-Based Implementation of Control Strategies

Measurements have been conducted at laboratory stand consisting of 3-phase APF prototypes. The APF system has been designed on the basis of voltage inverter using MOSFET IRFM460 transistors ($I_{Dmax} = 20$ A, $V_{DSmax} = 500$ V, $R_{DS} = 0.27$ Ω). Supply voltage is less than line voltage. Selected system parameters are as follows:

- Network voltage: $|V|_{a,b,c} = 63$ V
- Source inductance: $L_s = 0.8$ mH
- Inductance of compensating branch: $L_k = 1.2$ mH

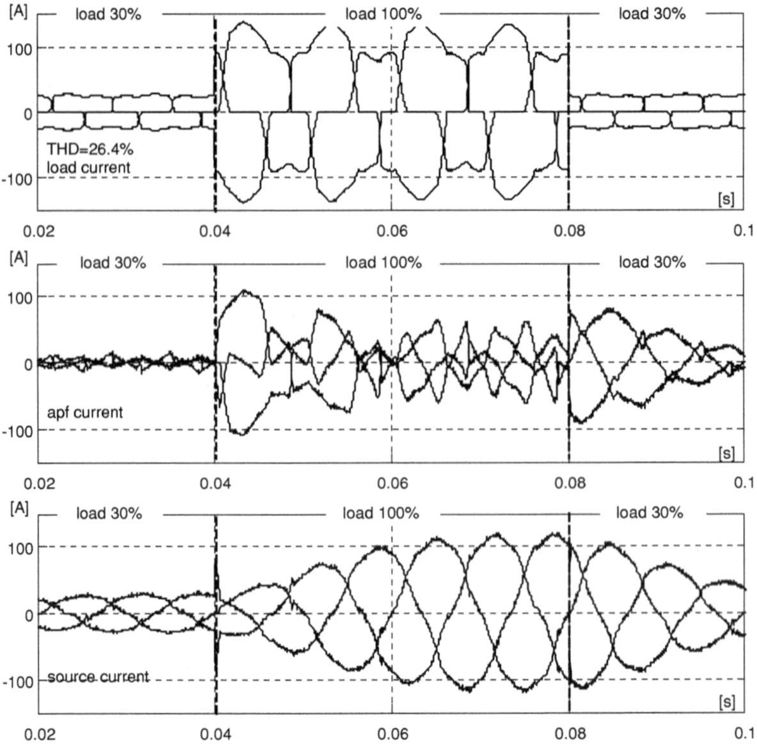

Fig. 3.60 Simulation results, THD coefficients relate to phase *a*

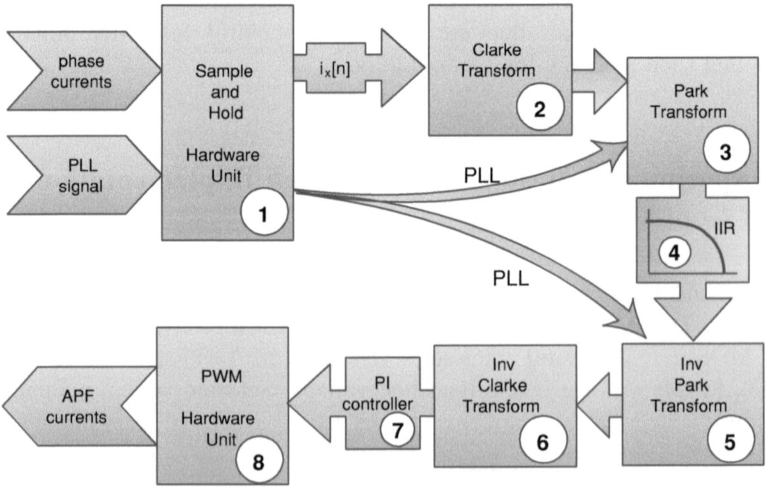

Fig. 3.61 Block diagram of control algorithm using synchronous reference frame theory

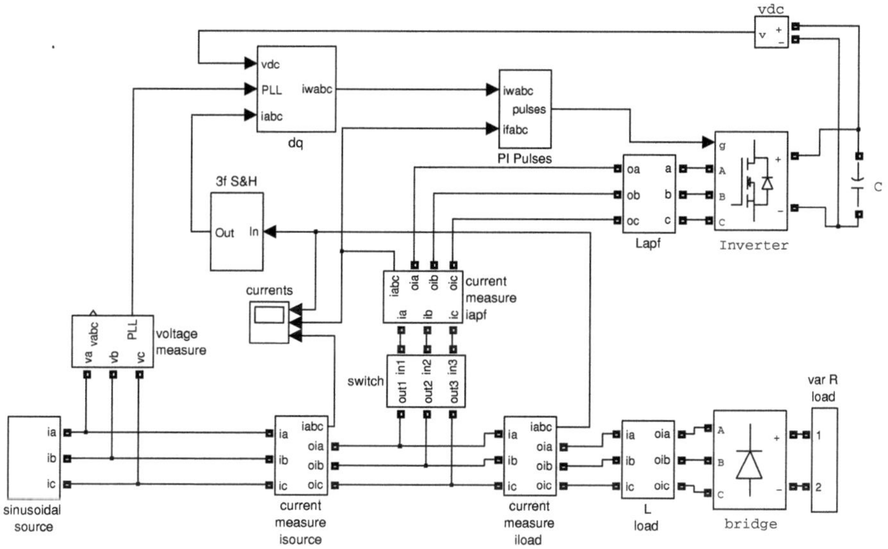

Fig. 3.62 Block diagram of the model elaborated in Matlab/Simulink environment

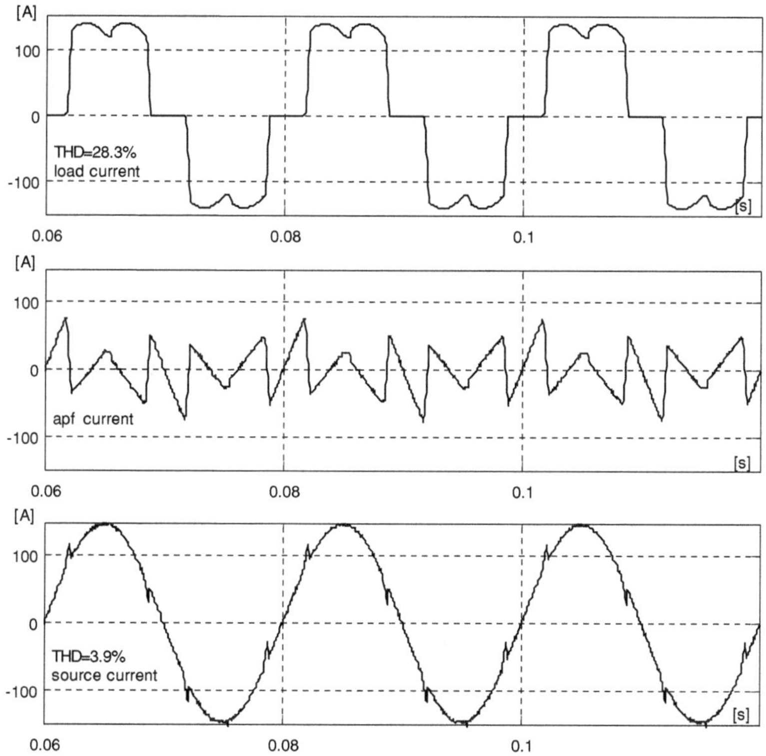

Fig. 3.63 Simulation results, phase *a*

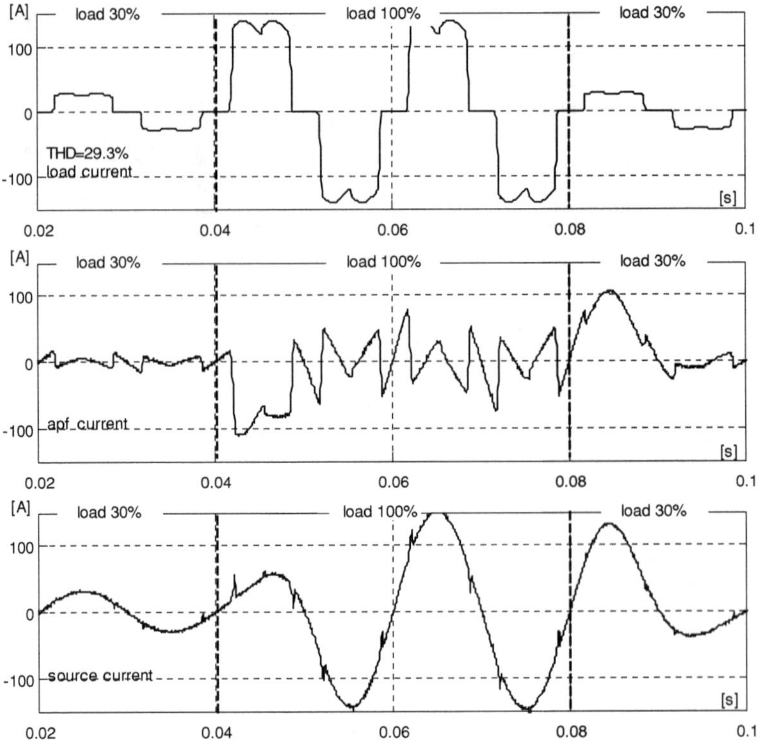

Fig. 3.64 Simulation results, phase *a*

- Capacitance of supply condenser: $C_{dc} = 4{,}300\ \mu F$
- V_{dc} link voltage: $V_{dcREF} = 200$ V
- Switching frequency: $f_{PWM} = 25$ kHz

Control circuit of EFA and HEFA prototypes has been implemented on the basis of starter kit using signal microcontroller TMS320F28335 manufactured by Texas Instruments (Fig. 3.71).

3.3.1 Digital Signal Controllers

Digital signal controller is a functional association of traditional industrial controller with computational core of digital signal processor. This helps to take full advantage of the best features of both solutions. Signal controller is able to cooperate with an extensive range of peripheral devices via different in-built ports and interfaces such as GPIO, CAN, McBSP, SPI, PWM. DSP core makes it possible to implement:

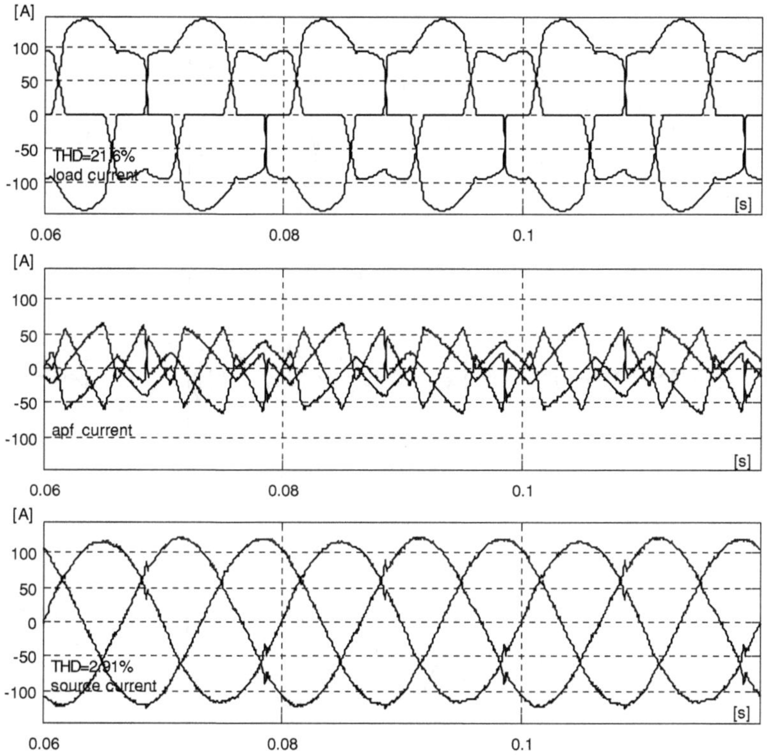

Fig. 3.65 Simulation results, THD coefficient relates to phase *a*

- One-cycle multiplication and accumulation cycles (one of the commonest operations in digital processing is sum of the products)
- Hardware shift units
- Special address modes (e.g. circular)
- Parallel ALUs
- Pipelining

This type of architecture makes it possible to implement complex signal processing algorithms and control processes quickly. That is why they are extensively used in many different industry branches and in particular in drive control systems, power conversion systems or measurement signal processing. Among the most important manufacturers of these systems are Texas Instrument, Free scale and Microchip.

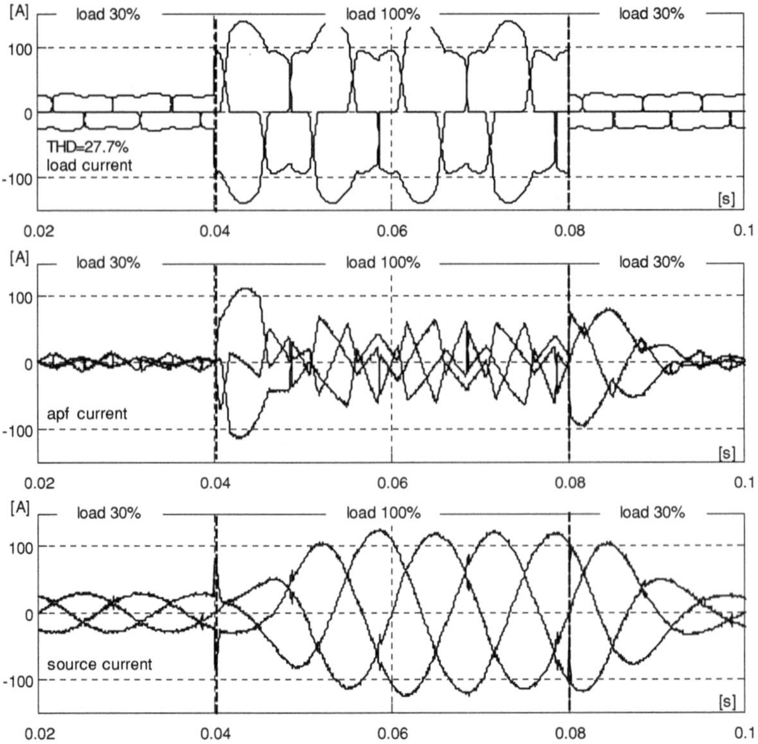

Fig. 3.66 Simulation results, THD coefficient relates to phase *a*

3.3.1.1 Floating Point DSP Controller

Most of present signal microprocessors uses fixed point arithmetic. Such systems are cheaper to manufacture, while ensuring correct implementation of most algorithms. When high precision is required, floating point systems are used. In APF control circuits digital low-pass IIR filters are often applied. The cut-off frequency of these filters is usually less than 100 Hz and with sampling rate of 20 kHz fixed point filter's parameters are saddled with significant quantification error. Therefore, we decided to apply TMS320F28335 signal microcontroller to implement algorithms described in previous sections. Texas Instruments company is the sole manufacturer of floating point DSP controllers. Its experience of many years standing has helped to work out this system, providing really huge possibilities and performance 300 Mflops. This system belongs to C2000 family with most up-to-date F28x family core. Figure 3.72 shows in graphical form the development of processors belonging to this family (illustration is based on data from Texas Instrument web page).

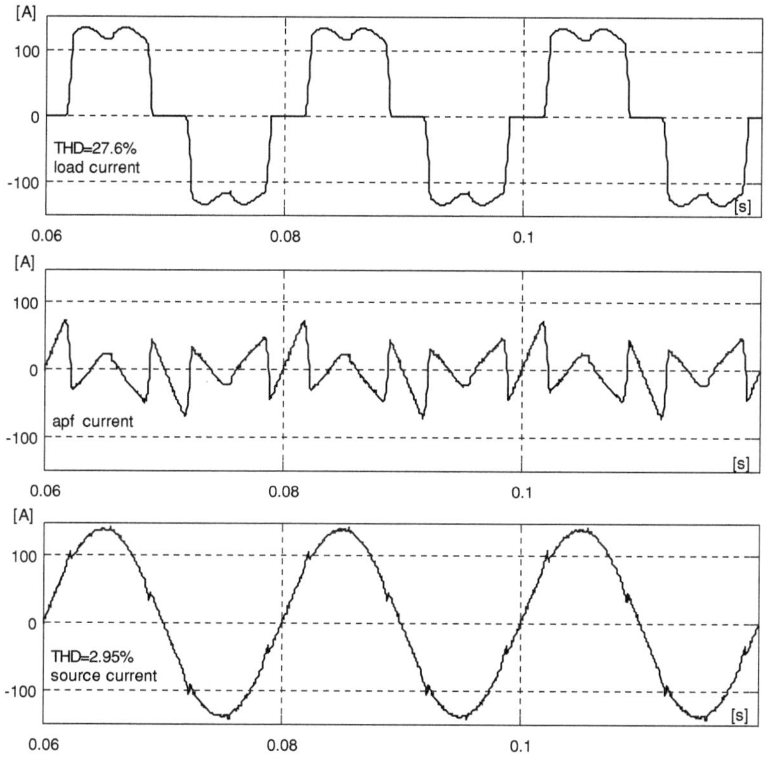

Fig. 3.67 Simulation results, phase *a*

Some characteristic features of this processor are listed below:

- Instruction cycle time 6.67 ns
- 256 kW (kilo words) Flash memory, 34 kW SARAM memory
- Single precision floating point unit compatible with IEEE-754
- Six-channel DMA controller
- 88 general purpose ports (GPIO)
- Interrupt Management Unit (PIE)
- 18 PWM circuits, 6 high precision PWM units
- Three 32-bit timers
- Two CAN units
- Two fast serial ports (McBSP)
- 16 channels of analog–digital converter (12 bit)
- Six DMA channels
- Compatibility with real-time hardware debugger via JTAG interface

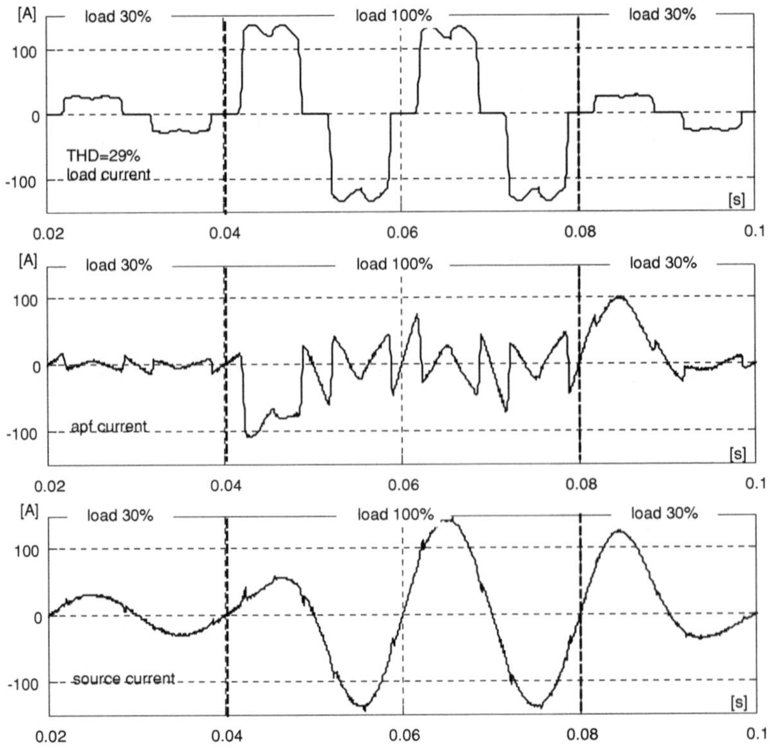

Fig. 3.68 Simulation results, phase *a*

3.3.1.2 Programme Developing Tools

DSP high computing power makes it possible to use high-level programming language C++. Only if very highly effective algorithms are necessary, a low level language (assembler) is used. This facilitates fast elaboration of software and, at the same time, costs are lowered. It also helps to be able to use algorithm libraries provided by microcontroller manufacturers. Usually the software is created in some provided integrated development environment (IDE). In case of Texas Instruments DSPs this is called Code Composer Studio. This environment is a highly elaborate application supervising operation and software of all Texas Instruments DSPs as well as ARM and MSP430 family processors of the same manufacturer. It simplifies and accelerates the creation of most complex applications. Additionally it provides the possibility of simulation testing as well as real-time testing via hardware interface JTAG. User is provided with extensive help system, possibility of implementing GEL scripts and using additional plugs worked out by external companies. With the arrival of new version of this environment, i.e. 4.0, user interface underwent a complete change. The environment has been the sole creation of TI until version 3.3 (Fig. 3.73), and since version 4.0

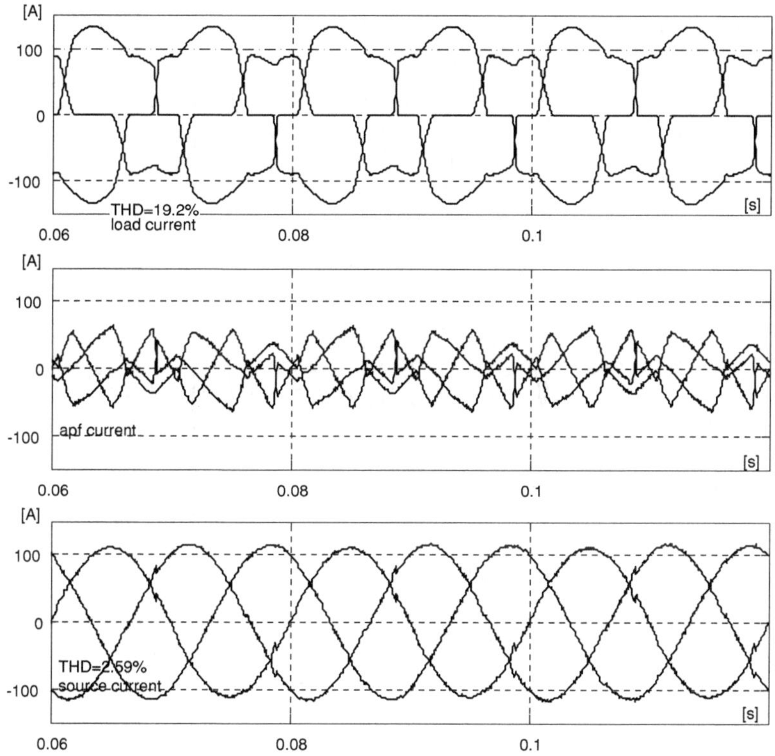

Fig. 3.69 Simulation results, THD coefficient relates to phase a

was introduced (Fig. 3.74) universal free programming environment provided by Eclipse company has been used. This facilitated significant improvement of CCS possibilities as well as improved user-friendly potential.

3.3.2 Performance Analysis of Control Strategies

At present author of this chapter has worked out all presented algorithms using time domain. Currently algorithms using frequency domain are elaborated. Great problem is posed by optimisation of FFT algorithms implementation and synchronisation of result with the phase current time waveform. The results will be presented in future publications. Control algorithms have been implemented with C++ language in Code Composer Studio v 3.3 environment described in previous section. All presented algorithms use PWM interrupt management procedure generated at 25 kHz frequency. High efficiency of F28335 processor ensures real-time implementation of these algorithms (see Fig. 3.1 for explanation of "real-time" concept). Figure 3.75 provides statistics on different algorithms efficiency.

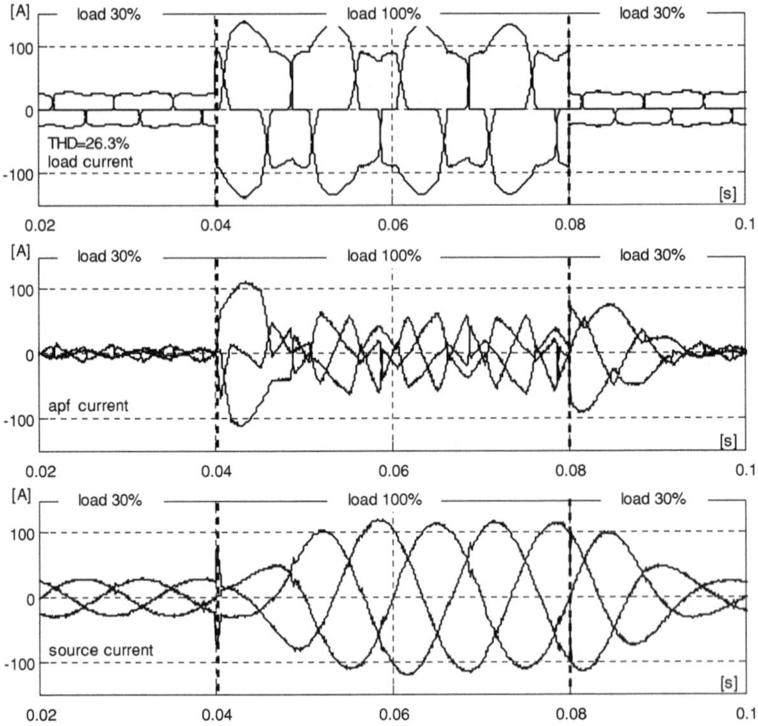

Fig. 3.70 Simulation results, THD coefficient relates to phase *a*

Number of clock cycles (Fig. 3.75) during which calculations are run are shown here.

For comparison sake, Fig. 3.76 shows times of implementation of complete time algorithms as opposed to one 128 point FFT (the frequency algorithm must contain 6 complete FFTs and 3 inverse FTs).

You can see that computing power needed to implement algorithms in the frequency domain is much higher than time algorithms. Therefore, the implementation of this algorithm using the conventional control systems used in industry is not possible in real time (Fig. 3.1). However, the implementation of several periodic delay may cause problems with synchronising waveform currents. The author's research can be concluded that the frequency algorithms, for example, based on the theory of the CPC, despite a better interpretation of the phenomena occurring in the system are not suitable for general use in industrial solutions of APFs.

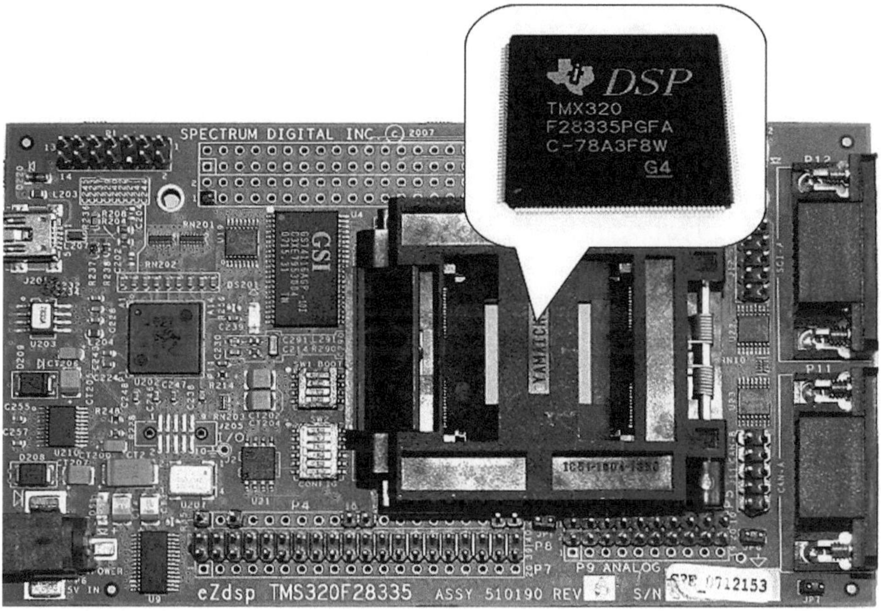

Fig. 3.71 DSP starter kit with TMS320F28335 digital signal controller

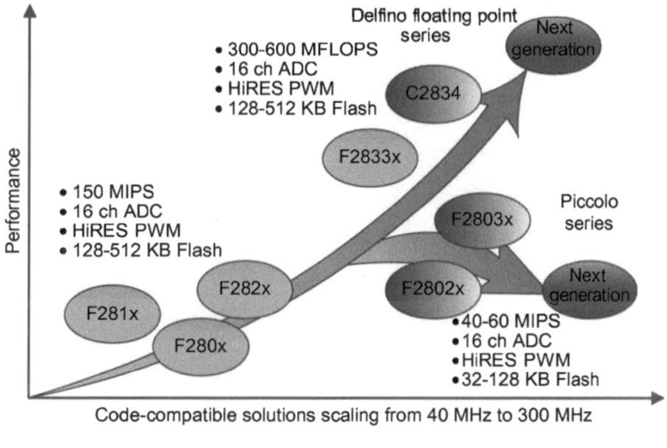

Fig. 3.72 Texas instruments processor family with F28x core

3.3.3 Results and Properties Comparison

Table 3.2 gives comparison and application range of different algorithms. The results have been obtained on the basis of simulations, lab tests and theoretical analysis.

Presented algorithms have been implemented in APF model and oscillograms shown in Figs. 3.77, 3.78, 3.79, 3.80, 3.81, 3.82 have been obtained. Figure 3.77

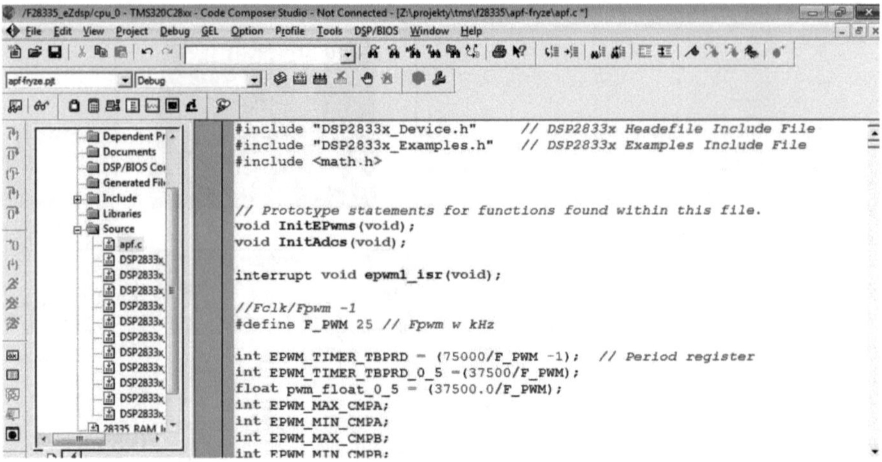

Fig. 3.73 Code composer studio, version 3.3

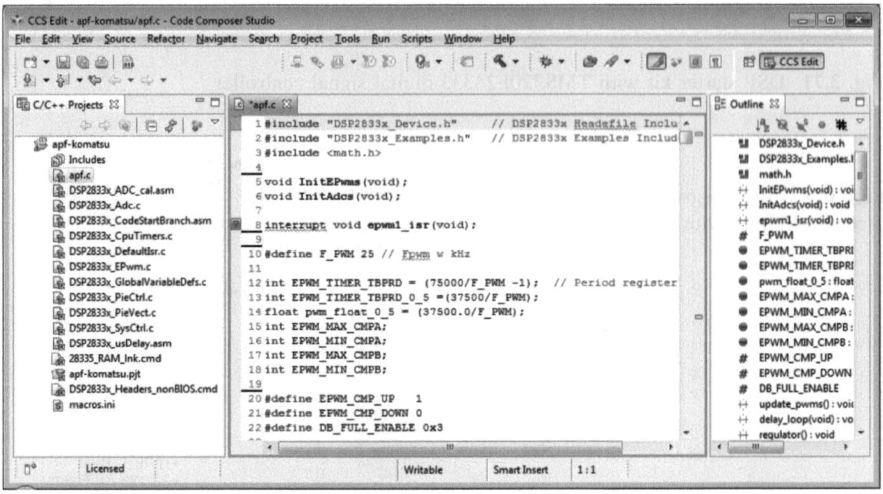

Fig. 3.74 Code composer studio, version 5.0

shows source voltage and current, load current, APF current in phase *a* in compensation circuit utilising Fryze's theory.

Figure 3.78 presents oscillograms of load current, APF current and phase *a* source current in compensation circuit utilising *pq* theory.

Figure 3.79 shows oscillograms of source voltage and current, load current and APF current in compensation circuit utilising extension *pq* theory.

Figure 3.80 shows source current, load current, APF current in phase *a* in compensation circuit using *dq* theory.

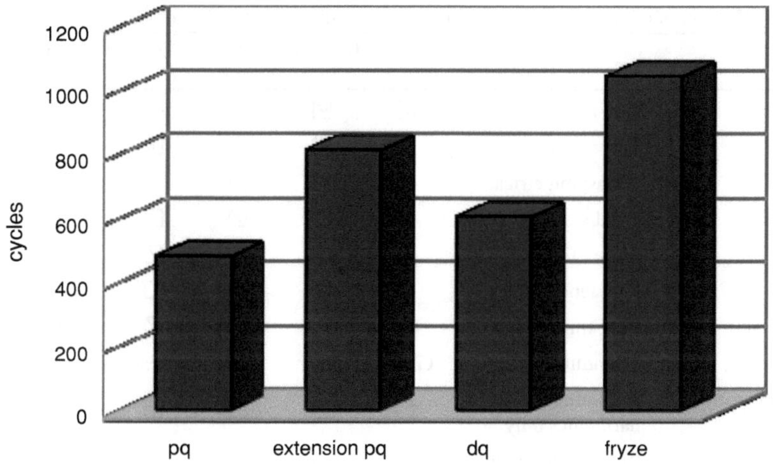

Fig. 3.75 Number of processor cycles for different algorithms, during which compensating currents are calculated

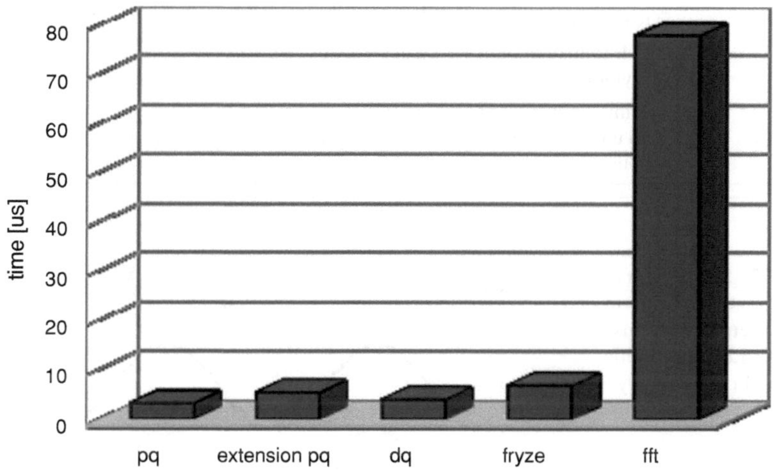

Fig. 3.76 Comparison of cycle time of time algorithms with cycle time of 128 point FFT algorithm

Figure 3.81 shows examples of oscillograms of transient state (unit step change of load power) in compensation circuit using Fryze's theory.

Figure 3.82 shows load current, source current and capacitor voltage (*PI* controller) waveforms in case of unit step change in load power.

Table 3.2 Comparison of selected features of presented control algorithms

		Fryze	Budeanu	CPC	p-q	Extension p-q	d-q
Supply	One-phase	☑	☑	☑	☐	☐	☐
	Three-phase	☑	☑	☑	☑	☑	☑
	- asymmetrical	☑	☐	☐	☐	☑	☐
	- distorted	☑	☐	☑	☐	☐	☑
Domain	Time	☑	☐	☐	☑	☑	☑
	Frequency	☐	☑	☑	☐	☐	☐
Efficiency	Real-time	☑	☐	☐	☑	☑	☑
	Dynamics	Good	poor	poor	best	good	best
Compensation properties	a) Elimination of harmonics only	☐	☑	☐	☑	☐	☐
	b) Elimination of selected harmonics	☐	☑	☐	☐	☐	☑
	c) Elimination of negative-sequence component	☐	☑	☐	☑	☐	☐
	d) Compensation of reactive power	☐	☑	☑	☑	☐	☐
	e) Elimination of asymmetry	☐	☑	☑	☐	☐	☐
	Joint elimination of all undesirable components (a-e)	☑	☑	☑	☑	☑	☑

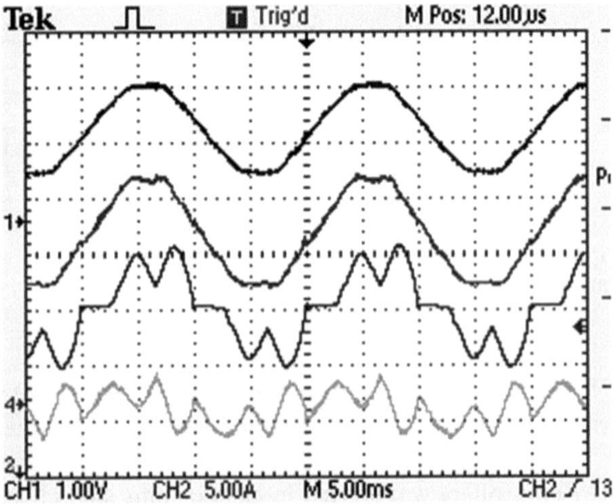

Fig. 3.77 Source voltage and current, load current and APF current waveforms; APF controlled by algorithms utilising Fryze's theory

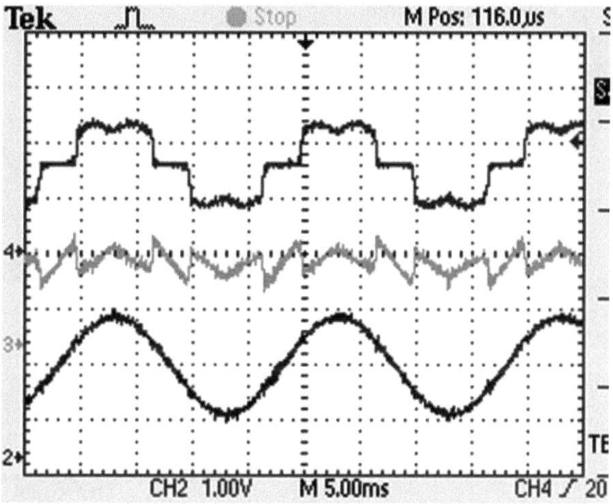

Fig. 3.78 Load current, source current and APF current waveforms; APF controlled by algorithm utilising *pq* theory

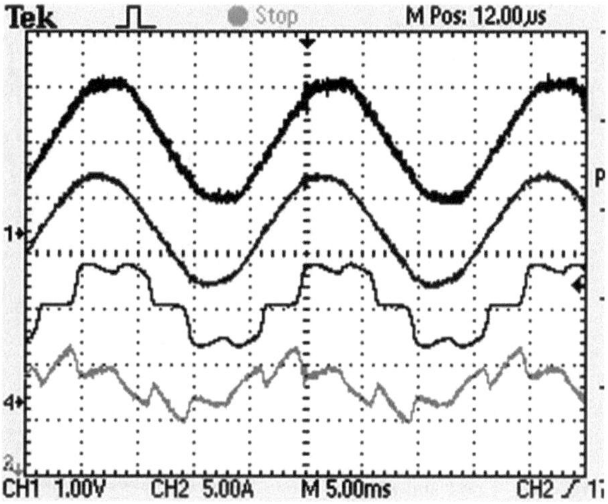

Fig. 3.79 Source current and voltage, load current and APF current waveforms; APF controlled by algorithm utilising extension *pq* theory

3.3.4 Concluding Remarks

Modelling the control circuit in Matlab/Simulink environment has provided elasticity when control algorithm parameters were selected; on the other hand, quite precise representation of APF real operating conditions has been possible (due to application of digital filters and accounting for delays in acquisition and data processing circuits). The prepared algorithm models have subsequently been implemented in lab prototype of APF system. The principal control element of this APF is TMS320F28335 signal microcontroller (DSP). Presented results demonstrate that the algorithms have been correctly implemented.

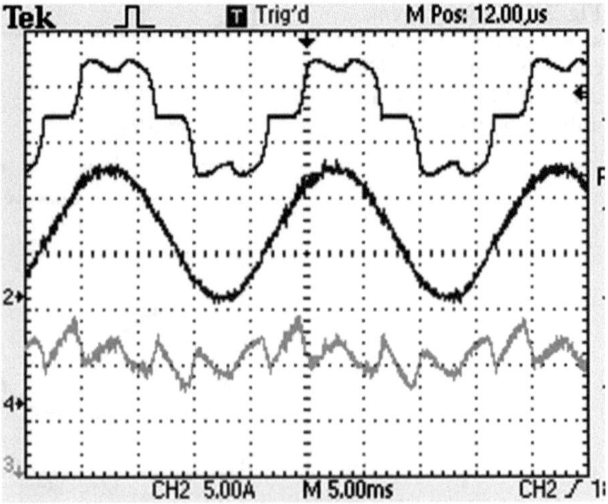

Fig. 3.80 Source current, load current and APF current waveforms; APF controlled by algorithm utilising *dq* theory

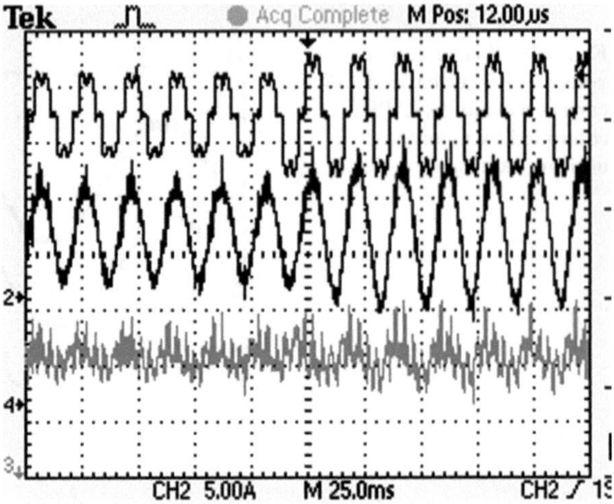

Fig. 3.81 Load current, source current and APF current waveforms for exemplary transient state; APF control algorithm utilising Fryze's theory

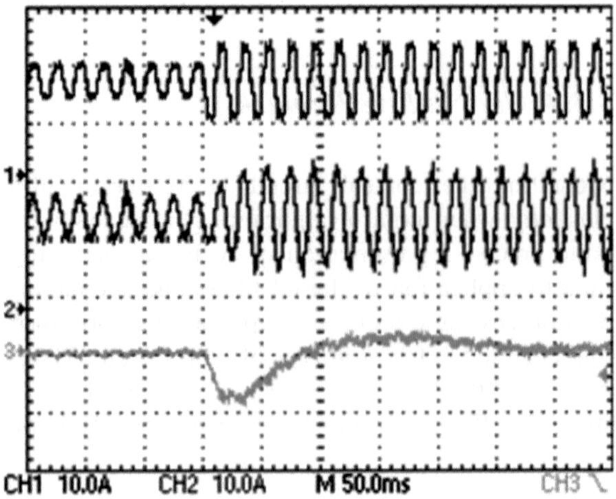

Fig. 3.82 Load and source current, V_{dc} capacitor voltage waveforms during change in load

References

1. Kowalski Z (2007) Power quality (in Polish). Technical University of Łódź, Łódź
2. Maciążek M, Pasko M (2009) Higher harmonics elimination methods in power systems (in Polish). In: PES conference, pp 9–14
3. Maciążek M, Pasko M (2007) Numerical algorithms for the improvement of working conditions of voltage sources (in Polish). Silesian University of Technology Press, Gliwice
4. Piróg S (1998) Power electronics. The negative impact of power electronic systems for energy sources and selected methods of their reducing (in Polish). AGH, Kraków
5. Rivas D, Moran L, Dixon JW, Espinoza JR (2003) Improving passive filter compensation performance with active techniques. IEEE Trans Ind Electron 50(1):161–170
6. Siwczyński M (2000) The theory of universal global passive compensation circuits. JUEE 6:77–86
7. Strzelecki R, Supronowicz H (2000) Power factor in AC power systems and methods of its improvement (in Polish). Warsaw University of Technology, Warszawa
8. Pasko M, Dębowski K (2002) Summarization of three-phase and multi-phase systems powered from distorted periodic voltage sources (in Polish). Silesian University of Technology Press, Gliwice
9. Afonso J, Couto C, Martins J (2000) Active filters with control based on the p-q theory. IEEE Ind Electron Newsl 47(3):5–10
10. Fukuda S, Furukawa Y, Kamiya H (2002) An adaptive current control technique for active filters. Power Convers Conf 2:789–794
11. Gawlik WHM (2003) Time domain modelling of active filters for harmonic compensation. IEEE Bologna Power Tech
12. Maciążek M, Pasko M (2009) Control of active power filters using instantaneous power theory (pq) (in Polish). Scientific papers of the Silesian University of Technology, Elektryka, vol 182, pp 69–87
13. Mikołajuk K, Toboła A (2009) Sliding mode observer for active power filters. Przegląd Elektrotechniczny 4:37–40
14. Mikołajuk T, Toboła A (2010) Average time–varying models of active power filters. Przegląd Elektrotechniczny 1:53–55
15. Pasko M, Buła D (2007) Hybrid active power filters. Przegląd Elektrotechniczny 7(8):1–5 (in Polish)
16. Strzelecki R, Supronowicz H (1998) Filtration of harmonics in AC power networks (in Polish). Toruń
17. Watanabae EH, Aredes M (2000) Compensation of non-periodic currents using the instantaneous power theory. In: IEEE PES summer meeting, pp 994–998
18. Woo-Cheol L, Taeck-Kie L, Dong-Seok H (2002) A three-phase parallel active power filter operating with PCC voltage compensation with consideration for an unbalanced load. Power Electron IEEE Trans 17(5):807–814
19. Akagi H, Watanabe EH, Aredes M (2007) Instantaneous power theory and applications to power conditioning. Wiley, New Jersey
20. Strzelecki R (1997) Using of instantaneous power theory to control of active power filters (in Polish). JUEE 3:65–74
21. Fryze S (1931) Active, reactive and apparent power in circuits with nonsinusoidal voltages and currents (in Polish). Przegląd Elektrotechniczny, 7: 193–203, 8: 225–234, 22: 673–676
22. Hanzelka Z (1991) Control of active power filter using vector instantaneous power theory (in Polish). In: Power quality conference, pp 111–117
23. Peng FZ, Lai JS (1996) Reactive power and harmonic compensation based on the generalized instantaneous reactive power theory for three-phase power systems. In: Proceedings of the IEEE 7th international conference on harmonics and quality of power, pp 83–89

24. Akagi H, Kanazawa Y, Nabae A (1984) Instantaneous reactive power compensators comprising switching devices without energy storage components. IEEE Trans Ind Appl 1A–20(3):625–630
25. Komatsu Y, Kawabata T (1995) A control method of active power filter where system voltage contains negative-phase-sequence component or zero-phase-sequence component. PEDS Conf 2:583–586
26. Komatsu Y, Kawabata T (1999) A control method for the active power filter in unsymmetrical voltage systems. Int J Electron 86(10):1249–1260
27. Komatsu Y, Kawabata T (1995) A control method for the active power filter in un-symmetrical voltage systems. In: Proceedings of EPE'95, pp 1.904–1.907
28. Maciążek M, Pasko M (2008) Control of three phase active filter using instantaneous power theory. In: IC-SPETO Conference, pp 303–308 (in Polish)
29. Bhattacharya S, Divan DM, Banerjee B (1991) Synchronous frame harmonic isolator using active seriesfilter. EPE Conf 3:3030–3035
30. Valouch V (1999) Extension p-q theory-based harmonic power compensation in unsymmetrical voltage systems. In: AMTEE Conference, pp C30–C33
31. Piróg S (2001) Active power filters and STATCOM fundamental harmonic of reactive current compensator controlled in dq rotating coordinates (in Polish). JUEE 7(1):91–105
32. Pasko M, Maciążek M, Buła D (2008) Significance of signal filters in control system of active power filter s (in Polish). Przegląd Elektrotechniczny 6:101–104

Chapter 4
Realization of a Digital Control Algorithm

Krzysztof Sozanski

Abstract Chapter four covers problems concerning the realization of digital control algorithms. The following realizations of digital control circuit are considered: digital signal processors, microprocessors, microcontrollers, and programmable digital circuits. All these solutions are compared. The general problems of analog signal acquisition for digital control circuit, such as sampling rate, number of bits, signal to noise ratio (SNR), anti-aliasing filter, bandwidth of signal, signal range are discussed too. Digital filter and filter bank are also considered, with special emphasis filters which are useful for active power filter (APF) control circuit, sliding discrete Fourier transform (DFT). Special attention is paid to improve APF dynamic range and using a non-causal solution it is possible to dump output stage dynamic distortion. For unpredictable loads a three-phase multirate APF with modified output inverter was designed. Finally, experimental test result of considered APF is presented and discussed.

4.1 Digital Control Circuit for Power Electronics

In recent years the field of power electronics has been dramatically expanded due to several factors. The most important one is the revolutionary growth of micro-electronics, which has influenced the development of analog and digital signal processors. Adding punch to this are the advances made in semiconductor fabrication technology which have made possible significant improvements in the

K. Sozanski (✉)
Institute of Electrical Engineering, University of Zielona Góra, 50 Podgórna Street, 65-246 Zielona Góra, Poland
e-mail: k.sozanski@iee.uz.zgora.pl

G. Benysek and M. Pasko (eds.), *Power Theories for Improved Power Quality*, Power Systems, DOI: 10.1007/978-1-4471-2786-4_4,

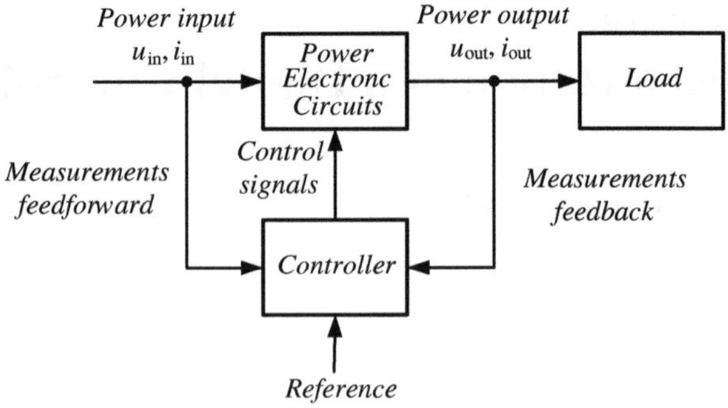

Fig. 4.1 Simplified diagram of exemplary power electronic system

current and voltage capabilities and the switching speeds of semiconductor power devices. In the 1970s, early power bipolar transistors were capable of carrying a few amperes and a few hundred volts with a switching frequency of around several thousands hertz. Today modern power transistors can carry hundreds of amperes and several thousand volts, with a switching frequency measured in millions of hertz. At the same time, the market for power electronics has significantly expanded. Power electronic circuits are now used everywhere, in industry, telecommunications, transportation, commerce, etc. They even exist in such modern popular devices as digital cameras, mobile phones, and portable media players. The background of power electronics is described by Mohan et al. [1], Erickson and Maksimovic [2], Bose [3], Trzynadlowski [4] and many others. A simplified diagram of an exemplary power electronic system is shown in Fig. 4.1. The system shows the conversion of electric power from input to load. The power input is typically from power line, electrochemical battery, solar cells or fuel cells. The input power can be DC or AC, single-, two-, three-phase, or more. The output power depends on the load. The power electronic circuit is monitored by a controller which compares the output power (feedback) and input power (feedforward) with the reference value to achieve the desired results. The controller can be designed to use either an analog or digital technique. Today the digital solution is most common, while the analog solution is reserved only for simple power electronic systems. The power flow through the electronic system may be reversible. The most common power electronic system component is the inverter.

Single- or three-phase inverters and even more numerous phases are frequently used parts of power electronic systems, in such devices as: AC and DC motor drives, uninterruptible power supplies, harmonic compensators, DC power supplies, controlled rectifiers, AC and DC power transmission systems, etc.

A simplified block diagram of an exemplary three-phase inverter with digital controller is shown in Fig. 4.2. The inverter consists of six power IGBT transistors Q_1, Q_2, Q_3, Q_4, Q_5 and Q_6, controlled by a driver with galvanic isolation. This galvanic isolation should be of low input-to-output capacitance and should be highly resistant to a high output voltage slew rate in the range 10–0.20 kV/μs. One of the inverter's legs consists of two transistors Q_1 and Q_4, which are connected to the load Z_{L1} through a *LC* low-pass filter (L_{f1}, C_{f1}) for suppressing

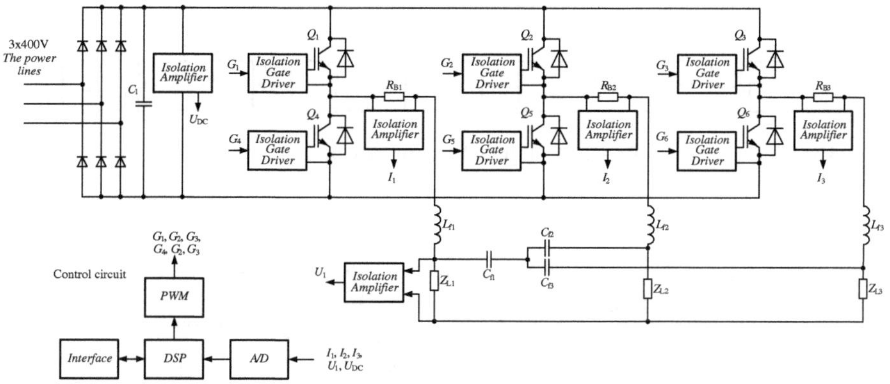

Fig. 4.2 Block diagram of an exemplary three-phase inverter with digital controller

pulse width modulation (PWM) components. The inverter is controlled by a Digital Signal Processor (DSP) which performs the control algorithm. The control DSP function can be realized using one of several devices: general-purpose microprocessors, microcontrollers, advanced microprocessors, and advanced microcontrollers, digital signal processors and programmable digital devices, etc. As with transistor control signals, analog signals representing currents and voltages are galvanically isolated too. The analog signals are converted into the digital form by an analog-to-digital (A/D) converter. The control algorithm monitors the output signal value of the transistors by pulses generated by the same kind of pulse modulator.

There are not many comprehensive publications dealing with digital control circuits for power electronic systems, though it is possible to find a lot of limited discussion in many conference papers. The background of digital control circuits for power electronics is described by Kazimierkowski et al. [5], and [6], Buso and Mattavelli [7]. The crucial first questions for digital control circuits involve:

- sampling rate;
- number of bits

4.1.1 Multirate Digital Control Circuit

Originally, the digital control algorithms were simply a representation of analog control circuits with a single sampling frequency, while later they turned into highly advanced and more complicated solutions. The block diagram of a classical digital control circuit is depicted in Fig. 4.3. This control circuit has a single analog input signal $x(t)$ and single analog output signal $y(t)$. Signal $x(t)$ is sampled at a frequency f_{s1} and the output signal is sampled at frequency f_{s3}. The control algorithm is

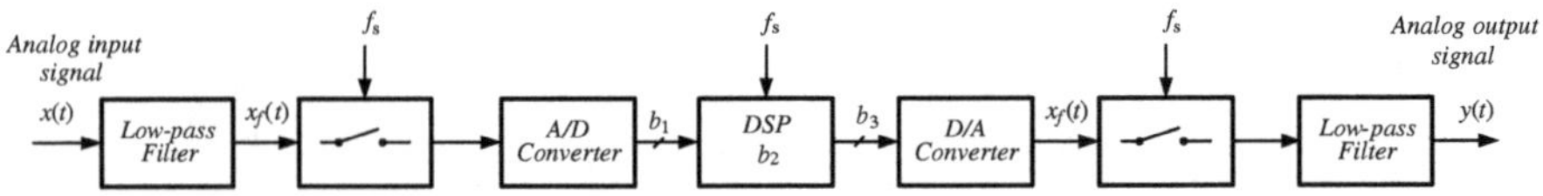

Fig. 4.3 Block diagram of a classical digital control circuit

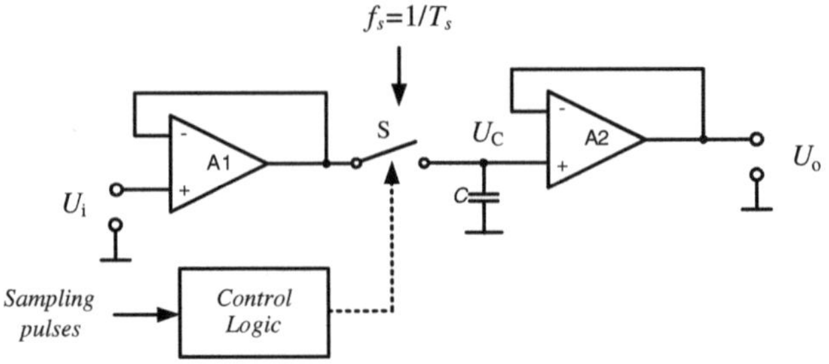

Fig. 4.4 A sampling circuit

performed by DSP at frequency f_{s3}. In the classical signal processing systems, all three frequencies are equal $f_s = f_{s1} = f_{s2} = f_{s2}$. Currently control circuits designed to increase signal dynamics use a few sampling frequencies. Digital circuits using different sampling rates are called multirate digital circuits. The multirate circuits are well described by Crochiere and Rabiner [8], Vaidyanathan [9], Fliege [10].

4.2 Signal Conditioning and Sampling

4.2.1 Analog Signal Sampling Rate

A continuous analog signal is sampled at discrete intervals, $T_s = 1/f_s$. The signal sampling frequency f_s (also called signal sampling speed) must be carefully chosen to ensure an accurate representation of the original analog signal. It is evident that the more samples are taken (faster signal sampling rates), the more accurate the digital representation is. Therefore if fewer samples are taken (lower sampling rates), a point is reached where critical information about the signal is actually lost. The discussion in this chapter is limited to periodic or uniform sampling. The exemplary sample and hold circuit is shown in Fig. 4.4 and an illustration of an analog sinusoidal signal sampling process is shown in Fig. 4.5.

The mathematical background of sampling theory was determined by Nyquist [11, 12] of Bell Telephone Laboratories. This original work was shortly supplemented by Hartley [13] and Whittaker. These papers formed the basis for the pulse code modulation (PCM) work to follow in the 1940s, and in 1948 Claude Shannon wrote his paper on communication theory [14]. Kotielnikov from the Soviet Union discovered independently of others sampling theory in 1933 [15]. Simply stated,

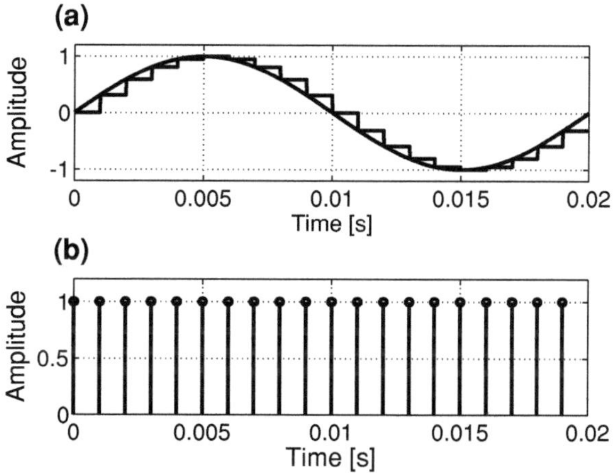

Fig. 4.5 Illustration of analog sinusoidal signal sampling process: **a** input signal and sampled signal, **b** sampling pulses

the sampling theory criteria require that the sampling frequency must be at least twice the highest frequency contained in the analog signal, or information about the analog signal will be lost. If the sampling frequency is less than twice the maximum analog signal frequency, a phenomenon known as aliasing will occur. This phenomenon is a source of spurious signals occurring within the anti-aliasing filter as a result of the high signal bandwidth available in today's ADCs. The main purpose of an anti-aliasing filter is to limit the input signal's bandwidth to eliminate high-frequency components. A rule of sampled data systems is that the input signal's spectrum gets folded around a frequency of one-half that of the sample clock. An ideal anti-aliasing filter would pass all signals in the band of interest and block all signals outside of that band. The quality of the anti-aliasing filter is the major factor in the signal-to-noise ratio (SNR). The SNR can be defined by the equation:

$$\mathrm{SNR_{dB}} = 10\log_{10}\left(\frac{P_{\mathrm{x}}}{P_{\mathrm{n}}}\right) \tag{4.1}$$

where: P_{x}—power of signal, P_{n}—power of noise.

Figure 4.6 shows spectra of a sampling signal process with aliasing occurrence. In the classical system for signal band 0 … f_{b}, the half sampling frequency $f_s/2$ is only a little bit higher than f_{b}, so the anti-aliasing has to have very sharp amplitude characteristics and a high duping factor in the stop-band. In such a solution the anti-aliasing filter is very complicated and expensive.

Currently, the development of integrated circuit (IC) manufacturing technology has led to the fact that fast digital circuits are freely available and cheap. Therefore, in modern systems the value of the sampling rate f_{s} can be much higher than f_{b} (oversampling), so the requirements for the anti-aliasing filter are much lower.

The background of digital signal processing useful for A/D and D/A conversion is described by Oppenheim and Schafer [16], Proakis and Manolakis [17], Lyons [18], Rabiner and Gold [19] and many others. Some interesting technical problems

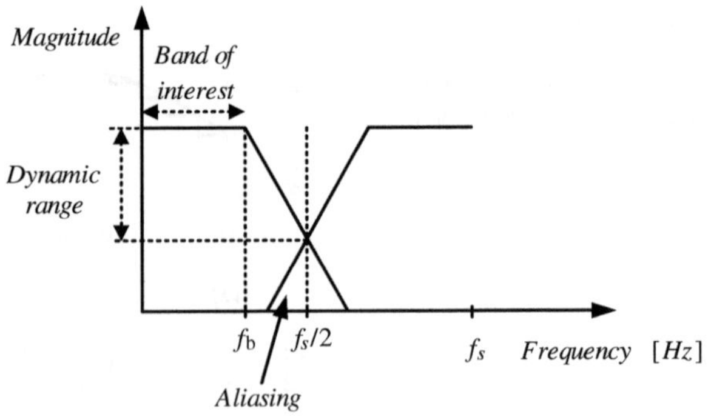

Fig. 4.6 An illustration of aliasing impact on the signal dynamics

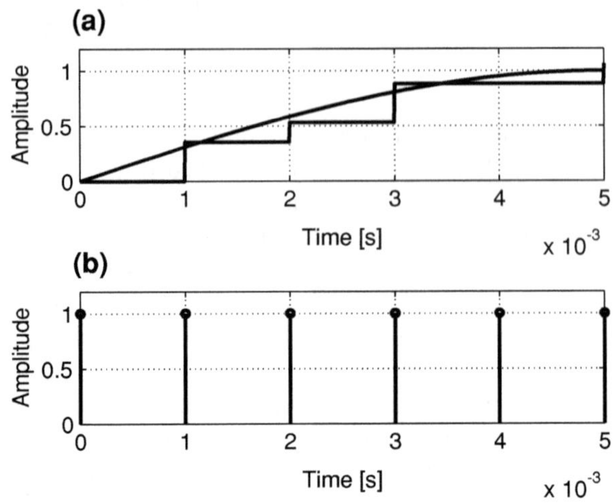

Fig. 4.7 Illustration of analog sinusoidal signal sampling and quantization process: **a** input signal and sampled signal, **b** sampling pulses

with A/D conversion are described in Data Translation [20, 21] and Analog Devices [22–24] (by Kester) technical reports and books.

4.2.2 Signal Quantization

Amplitude resolution of digital signal is limited to digital representation. The most commonly used is the b-bit fixed-point system, which can eliminate the excess digits by discarding them or by rounding off the number. An illustration of the sampling and quantization process of a sinusoidal analog signal is shown in Fig. 4.7.

The digital signal is a sequence of numbers in which each number is represented by a finite number of digits:

$$x(n) \overset{=}{\leftrightarrow} x(nT_{\mathrm{s}}), \quad -\infty < n < \infty \tag{4.2}$$

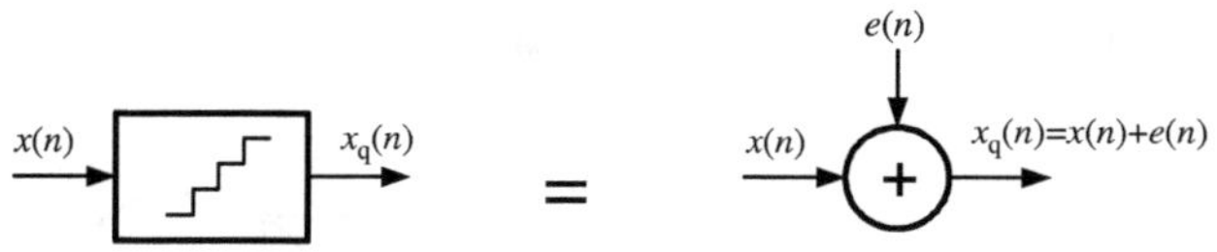

Fig. 4.8 The additive linear model of quantization process

where: n—number of samples, $x(n)$ discrete signal obtained by sampling of analog signal $x(t)$ every T_s period of time.

The amplitude of the signal corresponding to the last significant bit (LSB) is determined by the equation:

$$\Delta = \frac{A_p}{2^{b-1}} \tag{4.3}$$

where: A_p—maximum amplitude of converted signal, Δ is also called quantization step size and resolution.

The additive linear model of quantization process is shown in Fig. 4.8. The quantization error is defined by equation:

$$e_q(n) = x_q(n) - x(n) \tag{4.4}$$

$e_q(n)$ in rounding quantization process is limited to range of $-\Delta/2$ to $\Delta/2$, that is:

$$-\frac{\Delta}{2} \leq e_q(n) \leq \frac{\Delta}{2} \tag{4.5}$$

Signal quantization adds noise to the signal, which deteriorates the signal dynamic range. For a sinusoidal signal and quantization noise uniformly distributed, for fixed-pint b-bit system, signal to noise ratio can be determined:

$$\mathrm{SNR_{dB}} = 10\log_{10}\left(\frac{P_x}{P_n}\right) = 10\log_{10}\left(\frac{2}{3}2^{2b}\right) = 1.76 + 6.02b \tag{4.6}$$

4.2.3 Maximum Signal Frequency vs. Signal Acquisition Time

Signal acquisition time t_{aq} relates to A/D converters which use sample and hold (or track and hold) circuit on the input to acquire and hold (to a specified tolerance) the analog input signal, see Brannon and Barlow [25]. The signal acquisition time is the time required by the circuit to settle to its final value after it is paced in the hold mode. For an A/D converter without a sample and hold circuit on the input, the signal acquisition time is equal to the converter conversion time t_c. The only exception concerns flash converters with well matched comparators. An illustration of a sampling process is shown in Fig. 4.9, assuming that the amplitude of the

Fig. 4.9 Signal acquisition process

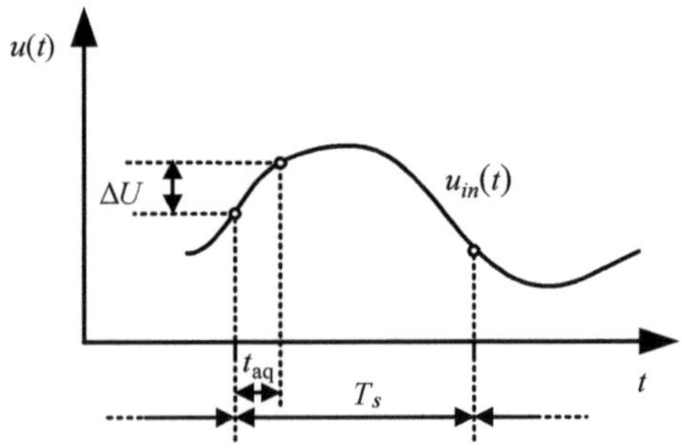

input signal in the acquisition process should not change more than half of the LSB of A/D converter. Assuming maximum signal change during sampling process:

$$\Delta U \le 0.5\Delta = 0.5\frac{A_{\mathrm{p}}}{2^{b-1}} = \frac{A_{\mathrm{p}}}{2^{\mathrm{b}}} \tag{4.7}$$

The analog input sinusoidal signal with amplitude A_{p} and frequency f:

$$u_{\mathrm{in}}(t) = A_{\mathrm{p}}\sin(2\pi f t) \tag{4.8}$$

The maximum speed of the signal change is determined by the equation:

$$\left.\frac{du_{\mathrm{in}}(t)}{dt}\right|_{\max} = 2\pi A_{\mathrm{p}} f \tag{4.9}$$

Assuming that:

$$t_{\mathrm{aq}} \ll 1/f \tag{4.10}$$

ΔU_{in} can be determined from:

$$\Delta U_{\mathrm{in}} = 2\pi A_{\mathrm{p}} f t_{\mathrm{aq}} \tag{4.11}$$

After straightforward algebraic manipulation is carried out we obtain an equation describing the maximum signal frequency:

$$\Delta U_{\mathrm{in}} \le \Delta U \tag{4.12a}$$

$$2\pi A_{\mathrm{p}} f t_{\mathrm{aq}} \le \frac{A_{\mathrm{p}}}{2^{\mathrm{b}}} \tag{4.12b}$$

$$f \le \frac{1}{2\pi 2^{b} t_{\mathrm{aq}}} \tag{4.12c}$$

Example 4.1 *Maximum Signal Frequency* For a number of bits $b = 16$ and acquisition time $t_{\mathrm{aq}} = 10$ ns, determined from the above inequality the maximum signal frequency is $f < 242.8$ Hz. As explained, this is one of the most important factors in the A/D conversion.

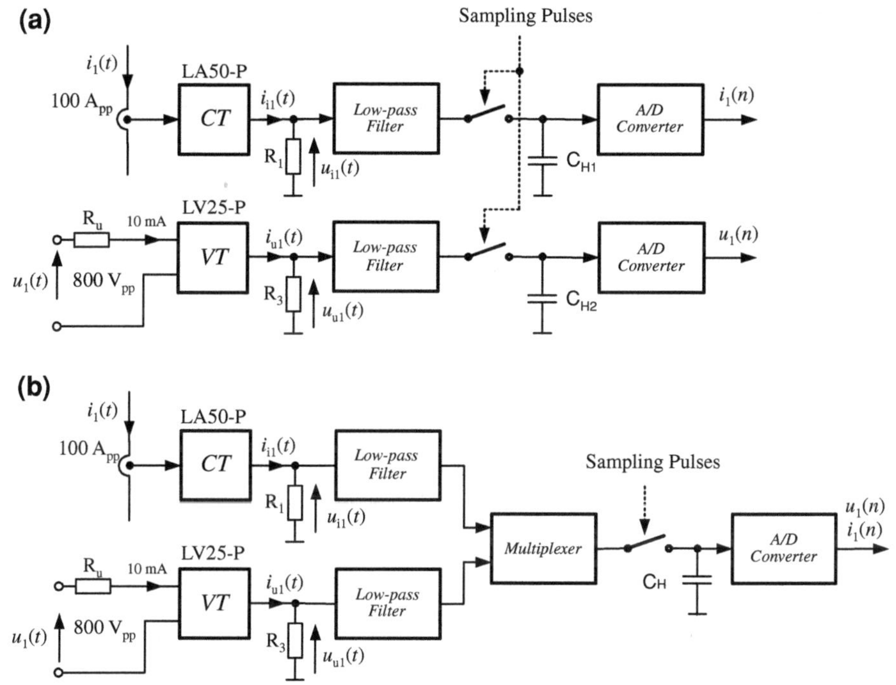

Fig. 4.10 Two-channel sampling circuit: **a** sequential sampling, **b** simultaneous sampling

4.2.4 Sampling in Multichannel System

A very important aspect of digital control circuits is the fact that the samples represent simultaneously sampled data or time-aligned data. The A/D converters with sampling circuit have two very common architectures. In the analog input design there are multiplexed and simultaneous sampling solutions. Simultaneous architectures use one A/D converter for many channels. A two-channel sequentially sampling analog-to-digital converter is depicted in Fig. 4.10a. Current $i_1(t)$ is converted to a current signal $i_{i1}(t)$ using galvanically isolated current transducer CT, in the same way a voltage signal is processed. In the next step the signals pass through the filter. Then the signals pass through anti-aliasing low-pass filters and are simultaneously sampled. After this the A/D converter process converts them to digital form.

Conversely, multiplexed architectures use one sampling circuit and A/D converter for every channel (Fig. 4.10b). The main disadvantage of a sequentially sampling A/D converter is the time error between channel samples. This time error is shown in Fig. 4.11, where two sample signals in time misalignment are shown. Discussion of simultaneous sampling versus sequential sampling is found in Data Translation reports [20, 21]. The best solution is a simultaneous sampling A/D

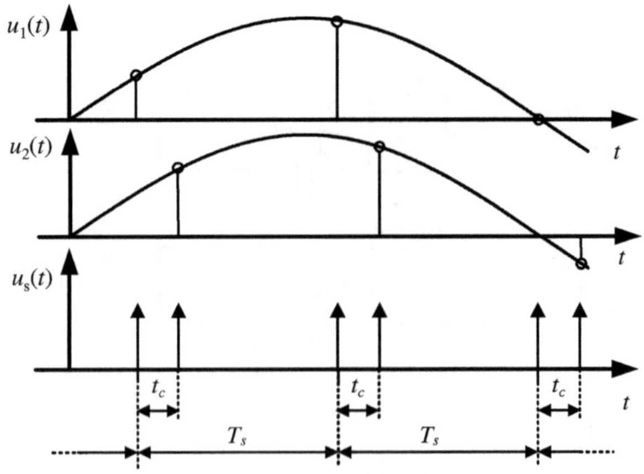

Fig. 4.11 Sequential sampling of two-channel sampling circuit

converter, however, if it cannot be used, the sequentially sampling A/D converter with time alignment has to be used. Benefits of simultaneous sampling compared to sequential sampling are:

- less jitter error;
- higher bandwidth of the system;
- less channel-to-channel crosstalk;
- less settling time

Amplitude and phase errors of sequential sampling A/D *conversion*
Sinusoidal input signals with unit amplitudes:

$$u_1(t) = \sin(2\pi ft), \quad u_2(t) = \sin(2\pi f(t + t_c)) \tag{4.13}$$

Difference of signals (Fig. 4.11):

$$\Delta U = \sin(2\pi ft) - \sin(2\pi f(t + t_c)) \tag{4.14a}$$

$$\Delta U = 2\cos\left(\frac{2\pi ft + 2\pi f(t + t_c)}{2}\right)\sin\left(\frac{2\pi ft - 2\pi f(t + t_c)}{2}\right) \tag{4.14b}$$

$$\Delta U = 2\cos(2\pi ft + \pi ft_c)\sin(-\pi ft_c) \tag{4.14c}$$

Maximum of ΔU:

$$\frac{d(\Delta U)}{dt} = -4\pi f\sin(2\pi ft + \pi ft_c)\sin(-\pi ft_c) \tag{4.15}$$

$$\frac{d(\Delta U)}{dt} = -4\pi f\sin(2\pi ft + \pi ft_c)\sin(-\pi ft_c) = 0. \tag{4.16}$$

So the derivative is zero when:

$$2\pi ft + \pi ft_c = 0 \tag{4.17a}$$

$$t = -\frac{t_c}{2} \tag{4.17b}$$

Maximum value of signal error can be calculated by formula:

$$\Delta U|_{t=-0.5t_c} = 2\cos\left(2\pi f\left(-\frac{t_c}{2}\right) + \pi ft_c\right)\sin(-\pi ft_c) \tag{4.18a}$$

$$\Delta U|_{t=-0.5t_c} = 2\cos(0)\sin(-\pi ft_c) \tag{4.18b}$$

$$\Delta U_{max} = 2\sin(-\pi ft_c) \tag{4.18c}$$

However, phase error can be determined from the equation:

$$\Delta\varphi = \frac{t_1}{T}360 - \frac{t_1 + t_c}{T}360 = 360t_c f \tag{4.19}$$

Example 4.2 *Maximum Signal Frequency for Sequential Sampling* For an A/D conversion time $t_c = 5$ μs and signal frequency $f = 50$ Hz it can be determined from the above equation that maximum signal error is $\Delta U = 1.57$ mV and phase error $\Delta\varphi = 0.09°$, and results for 50th harmonics ($f = 2{,}500$ Hz): $\Delta U = 39.26$ mV and $\Delta\varphi = 4.5°$. The result will be worse in multichannel systems with sequential sampling, where the A/D conversion time t_c for the last channel will be multiplied by the number of channels.

4.2.5 *Effective Number of Bits*

An interesting parameter of the digital signal is the effective number of bits (ENOB). This parameter takes account of all errors made during the conversion. Among the most important are: quantization noise, jitter, A/D converter noise, aliasing, integral and differential nonlinearity, channel crosstalk, channel-to-channel crosstalk, and other error sources (Fig. 4.12):

$$\text{ENOB} = \frac{\text{SNR}_{tot} - 1.76}{6.02} \tag{4.20}$$

where: SNR_{tot}—signal to noise ratio takes into account all the errors.

This parameter is especially important today when there is easy access to 18-bit A/D converters and by incompetent application it is easy to reduce the effective number of bits to 10–12. A major source of errors during the A/D conversion are digital transmission and clock signals of high-speed processors.

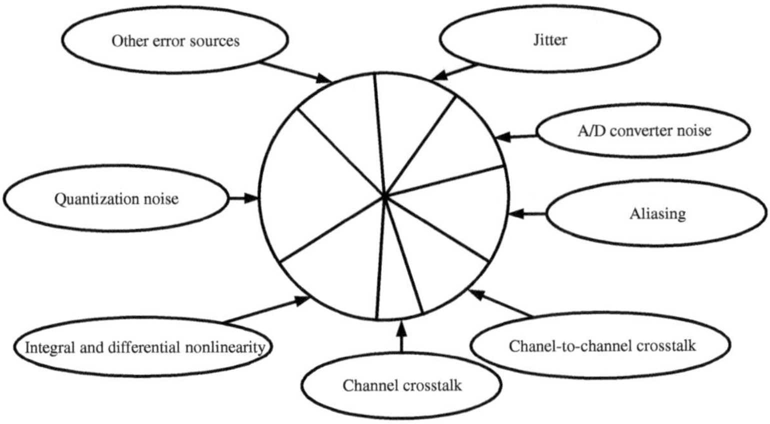

Fig. 4.12 Effective number of bits (ENOB)

Fig. 4.13 Spectra of sinusoidal signal: **a** coherent sampling, **b** no coherent sampling

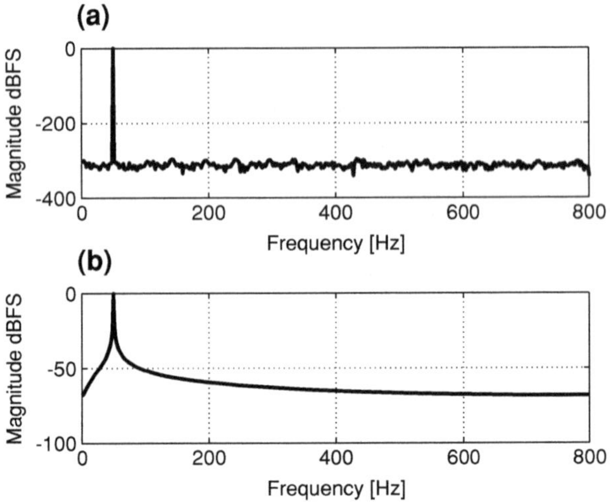

4.2.6 Synchronization of Sampling Process

The properties of a most of digital signal processing algorithms to a large extent depend on whether the processed signal is sampled coherently. For example, Fig. 4.13 shows the spectrum of the same signal with coherent sampling (Fig. 4.13a) and no coherent sampling (Fig. 4.13b). Therefore, in the author's opinion, for systems connected to the power network it is expedient to use synchronization. Figure 4.14 presents an analog synchronization circuit used by the author in active power filter (APF) control circuits. In this circuit the voltage of the three phases through isolation transformers are supplied to inputs of low-pass filters. Then, the signal from one phase is connected to the phase detector input of an analog phase lock loop (PLL). The PLL is generating sampling signal frequency

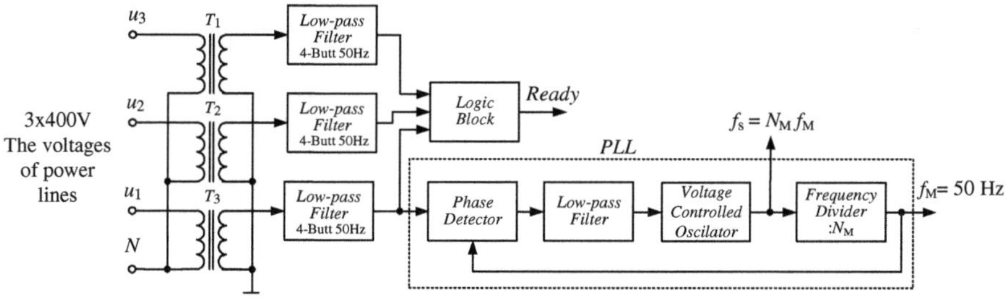

Fig. 4.14 Analog synchronization circuit with PLL circuit

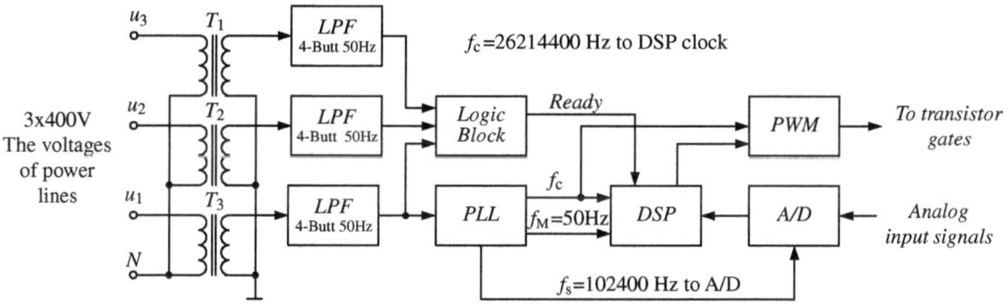

Fig. 4.15 Digital control system with full synchronization

$$f_s = N_M f_M \tag{4.21}$$

where: f_M—power line frequency, N_M—number of samples per power line period $T_M = 1/f_M$.

Another problem is the output of a power electronic device, usually this is a pulse width modulator generating pulse controlled output switches (transistors). When the modulation frequency is independent of the power line frequency, it becomes possible to generate low frequency components. This is a result of the formation of the beat frequency between the frequency of line and modulation. Therefore, the author considers that the whole control system should be synchronized. The block diagram of such a solution is depicted in Fig. 4.15. The DSP, PWM and A/D converter are synchronized by common PLL circuit.

4.3 Digital Signal Processors

Nowadays for DSP hardware it is possible to consider five main types of digital devices:

- general purpose Microprocessors (μP) and Microcontrollers (μC);
- fixed-point digital signal processors (DSP);
- floating-point DSP;

Table 4.1 Summary of DSP hardware implementation

	μP and μC	Fixed-point DSP	Floating-point DSP	FPGA	ASIC
Flexibility	Progra-mming	Progra-mming	Progra-mming	Progra-mming	None
Processing speed	Low-medium	High	High	High	High
Support for multiplication and accumulation	No-rare	Yes	Yes	Possible	Possible
Reliability	Medium–high	High	High	Medium	High
Resolution	Low	Low–medium	Medium–high	Low–medium	Medium
Additional peripheries: counters, PWMs, A/Ds	Yes	Possible	Possible	Yes	Yes
Power consumption	Low	Low	Medium–high	Low–medium	Low
Design time	Medium–long	Short	Short	Medium	Long
Design cost	Low–medium	Low	Low	Medium	High
Cost	Low–medium	Low–medium	Medium–high	Low–medium	Low

- programmable digital circuits;
- special-purpose devices such as application specific integrated circuits (ASIC)

Table 4.1 shows the main features of the processors. The main algorithms for DSP hardware can be described by equations: discrete convolution.

$$y(n) = \sum_{k=0}^{N-1} a(k) \cdot x(n-k) \tag{4.22}$$

correlation between random variables $w(n)$ and $x(n)$ at two different points in space or time:

$$y(n) = \sum_{k=0}^{N-1} w(k) \cdot x(n+k) \tag{4.23}$$

Discrete Fourier transform (DFT):

$$y(n) = \sum_{m=0}^{N-1} x(n)(\cos(2\pi nm/N) - j\sin(2\pi nm/N)) \tag{4.24}$$

The main task for DSP hardware can be described by equation:

$$y(n+1) = a(n) \cdot x(n) + y(n) \tag{4.25}$$

As for the A/D converter for fixed-point DSP number of bits determines the dynamic range of signals processing. The Fig. 4.16 shows the dynamic range of typical DSP.

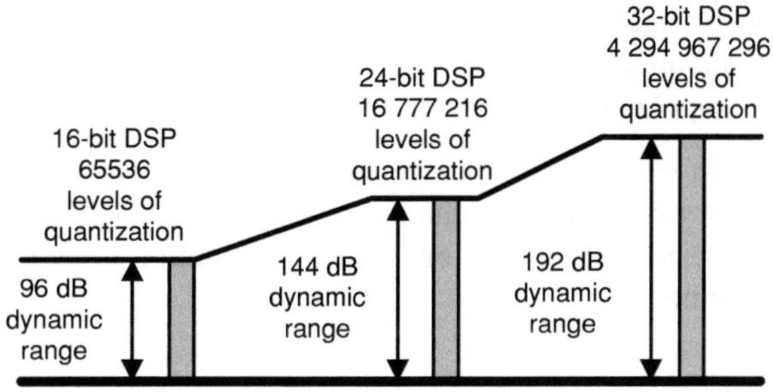

Fig. 4.16 Dynamic ranges of fixed-point digital signal processors

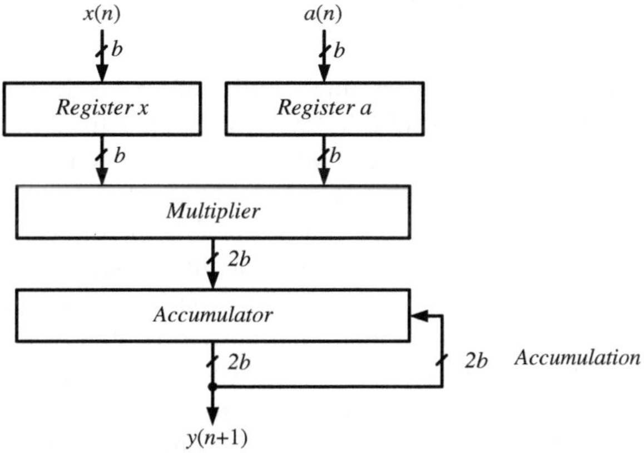

Fig. 4.17 Block diagram of multiplier with accumulator

The block diagram of multiplier with accumulator is shown in Fig. 4.17, input operands $x(n)$ and $a(n)$ have b-bit resolution, the product has $2b$-bit resolution. Accumulation is made using $2b$-bit resolution and finally it makes possible to calculate the basic Eq. 4.25 with better accuracy than b-bit:

$$\overbrace{y(n+1)}^{2b-\text{bit}} = \overbrace{\overbrace{a(n)}^{b-\text{bit}} \cdot \overbrace{x(n)}^{b-\text{bit}} + \overbrace{y(n)}^{2b-\text{bit}}}^{2b-\text{bit}}$$

It is a method to extend the dynamic range while keeping the cost of the system and its power consumption within reasonable limits. Such kind of multiplication with accumulation is a typical solution in DSP and it is rare in microprocessors and microcontrollers. It is possible to implement it by FGA and ASIC circuit.

On the market there are now very many different processors and it is very hard to choose the right one. In the author's opinion, the most important are the following features:

- single cycle operation;
- multiplier with extended resolution of accumulation;
- barrel shifter;
- hardware loops;

- circular addressing;
- parallel instruction set;
- support for fractional arithmetic

In contrast to what general-purpose processors provide, these features enable performing calculations quickly and with adequate precision. Table 4.2 shows the selected DSP useful for the implementation of control circuit for power electronic circuits.

4.3.1 TMS320F28335

One of the most interesting families of DSP is the family TMS320F28x, also called digital signal controller (DSC), from Texas Instruments [26]. A typical representative of this family is the TMS320F28335 system. It is a complete system with many useful features in a single silicon chip. Therefore, it is especially good for power electronics applications. The core of the processor consists of an IEEE-754 single-precision floating-point unit. For this application features especially useful are: 16-channel 12-bit ADC with 80-ns conversion rate and two sample and hold circuits, 18 PWM outputs, clock, and system control with dynamic PLL ratio changes, 256 K $\times$ 16 flash memory and 34 K $\times$ 16 SARAM memory. The instruction cycle of the processor is equal to 6.67 ns for a processor clock equal to 150 MHz. For simplifying the design process a ControlCARD module with TMS320F28335 is used. The ControlCARD is a small 100 pin DIMM (dual in line "memory module") style vertical plug-in board based on the F28335. These ControlCARDs have all the necessary circuits: clock, supply LDO, decoupling, pull-ups, etc., to provide reliable operation for the DSC device (Fig. 4.18). This reference design is very robust and is meant for operation in noisy electrical environments (especially important in power electronics), it includes the following features:

- all general purpose input/output (GPIO), ADC and other key signal routed to gold edge connector fingers;
- clamping diode protection at ADC input pins;
- anti-aliasing filter (noise filter) at ADC input pins;
- isolated UART communications.

The Docking Station is a small very basic mother board which accepts any member of the plug-in control card family. It provides the required 5 V power supply and gives the user access to all the GPIO and ADC signals. The module also has two small prototyping areas with an array of 0.1 " spaced plated through holes for soldering. The system for developing software is shown in Fig. 4.18. It consists of F28335 control card, CC28xxx docking station, and USB2000 Controller—JTAG emulator. Thanks to using the emulator it is possible to develop the software comfortably. The emulator has an access to the entire processor registers and memories and its use makes it possible to

Table 4.2 Selected DSP suitable for power electronics control circuit

Name	TMS320F283xx	TMS320C67x	SHARC	MC56F84xxx
Architecture	Harvard	Floating-point VLIW DSP core	Enhanced Harvard	Dual Harvard
Fixed-point	32-bit	32-bit	32-bit	32-bit
Floating-point	IEEE single-precision floating-point	IEEE double-precision floating-point	32/40-bit IEEE floating-point math	No
PWM	18 PWM, 6 high resolution (150 ps) PWM	Yes	Up to 16 PWM channels	Up to 24 PWM channels with 312 ps resolution
A/D	12-bit ADC with two sample and hold circuits	No	No	2 × 12-bit high speed
MAC	32 × 32 bit or dual 16 × 16 bit MAC	Two ALUs fixed-point, four ALUs, two multipliers floating -point and fixed-point	32-bit fixed-point multipliers with 64-bit product & 80-bit accumulation	32 × 32-bit with 32-bit or 64-bit result
Shifter	Barrel	Barrel	Barrel	32-bit arithmetic and logic multi-bit shifter
Signal processing performance	Up to 300 MHz operation with 600 MFLOPS	Up to 350 MHz/2,100 MFLOPS	450 MHz/2,700 MFLOPs	100 MIPS at 100 MHz
ROM	256 k × 16 on-chip flash memory	No	Up to 4 Mb	256 kB
RAM	34 k × 16	256 k	1–4 Mb	Up to 32 kB
Input/output	Up to 88 shared GPIO pins	Yes	Up to 16-bit	General purpose I/O
Special addressing modes	Circular addressing	Yes	Circular buffer addressing supported in hardware, 32 address pointers support 32 circular buffers	Parallel instruction set with unique DSP addressing modes
Hardware loop	Yes	Yes	Six nested levels of zero-overhead looping in hardware	Hardware DO and REP loops
Producer	Texas Instruments	Texas Instruments	Analog Devices	Freescale

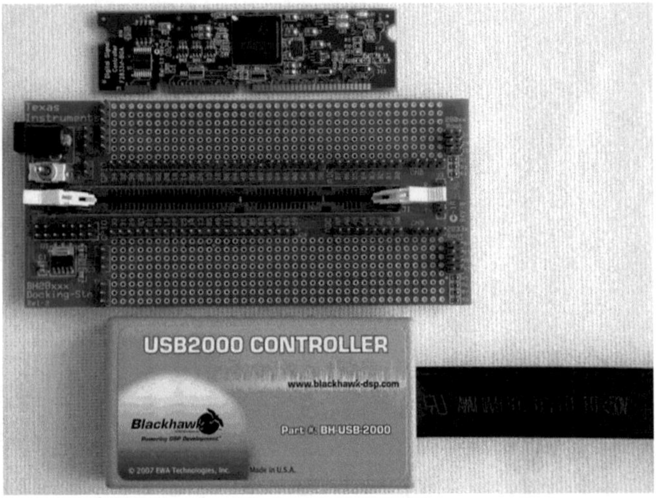

Fig. 4.18 System for developing software and hardware: F28335 control card, CC28xxx docking station, and USB2000 Controller

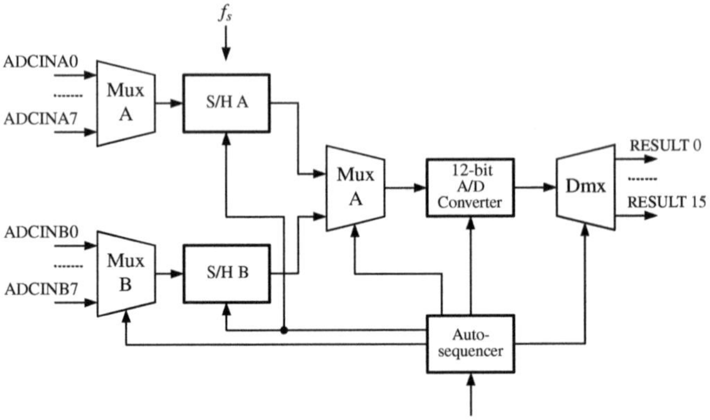

Fig. 4.19 Block diagram of TMS320F28335 A/D converterA/D converter

program internal flash memory. For software development Code Composer Studio v4 is used. Additionally Texas Instruments has prepared a lot of supporting tools, such as, Baseline Software Setup, DSP2833x Header Files, etc.

Simplified diagram of A/D converter is shown in Fig. 4.19. It consists of two sample and holds and one 12-bit A/D converter. The voltage input range is equal to 0…3 V. The converter input voltage U_{in} can be determined from the equation:

$$U_{\text{in}} = \frac{D(U_{\text{ref}+} - U_{\text{ref}-})}{2^b - 1} + U_{\text{ref}-} \tag{4.26}$$

where: D—converter digital output, U_{ref}—reference voltage, b—number of bits. For $U_{\text{ref}+} = 3$ V, $U_{\text{ref}-} = 0$ V and $b = 12$.

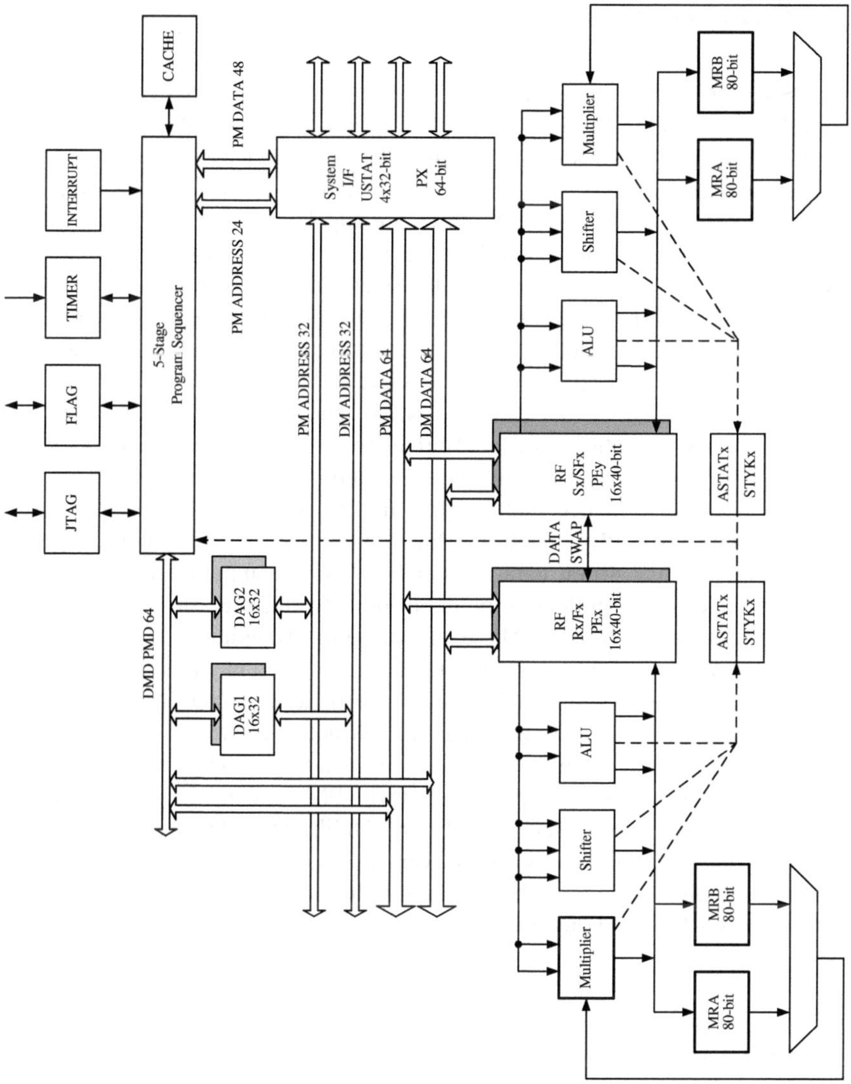

Fig. 4.20 Simplified block diagram of SHARC DSP core ADSP-21367/8/9

$$U_{\text{in}} = \frac{3D}{4,095} \tag{4.27}$$

4.3.2 SHARC DSP

In the author's opinion, a classic and the most programmer-friendly is the SHARC DSP family from Analog Devices. A block diagram of the SHARC core is shown in Fig. 4.20 [27]. The processor structure uses enhanced Harvard architecture, and consists of two sets of buses, one for Data Memory (DM) and a second for Program Memory (PM). PM and DM buses are capable of supporting 2 $\times$ 64-bit data transfers between memory and the core at every core processor cycle.

Address buses are controlled by two address calculators (arithmometers) DAG1 and DAG2. The DAGs are used for indirect addressing and implementing circular data buffers in hardware. Circular buffers allow efficient programing of delay lines and other data structures required in digital signal processing and are commonly used in digital filters (Eq. 4.22) and Fourier transform (Eq. 4.24). The two DAGs contain sufficient registers to allow the formation of up to 32 circular buffers (16 primary register sets, 16 secondary). The DAGs automatically handle address pointer wraparound, reduce overhead, increase performance, and simplify implementation. The SHARC has two computation units (PEx, PEy), each of which comprises: an ALU, multiplier with Barrel shifter, and 16 $\times$ 40-bit data register file. These computation units support IEEE 32-bit single-precision floating-point, 40-bit extended precision floating-point, and 32-bit fixed-point data formats. These units perform all operations in a single cycle. The three units within each processing element are arranged in parallel, maximizing computational throughput. Single multifunction instructions execute parallel ALU and multiplier operations. The processor includes an instruction cache that enables a three-bus operation for fetching an instruction and four data values. The cache is selective and only the instructions whose fetches conflict with PM bus data accesses are cached. The cache allows full-speed execution of core. Unlike other DSPs, SHARC has a programmer-friendly assembler, so it can very easily program the assembler code mixing with C language. The author considers that for standalone application the SHARC processor family should have a flash memory.

4.4 Digital Controller

4.4.1 Digital Controller for Shunt Active Power Filter

A simplified diagram of the conventional 75 kVA three-phase shunt active power filter with nonlinear load is depicted in Fig. 4.21. The APF was built by the University of Zielona Góra team [28], in which the author was involved in the

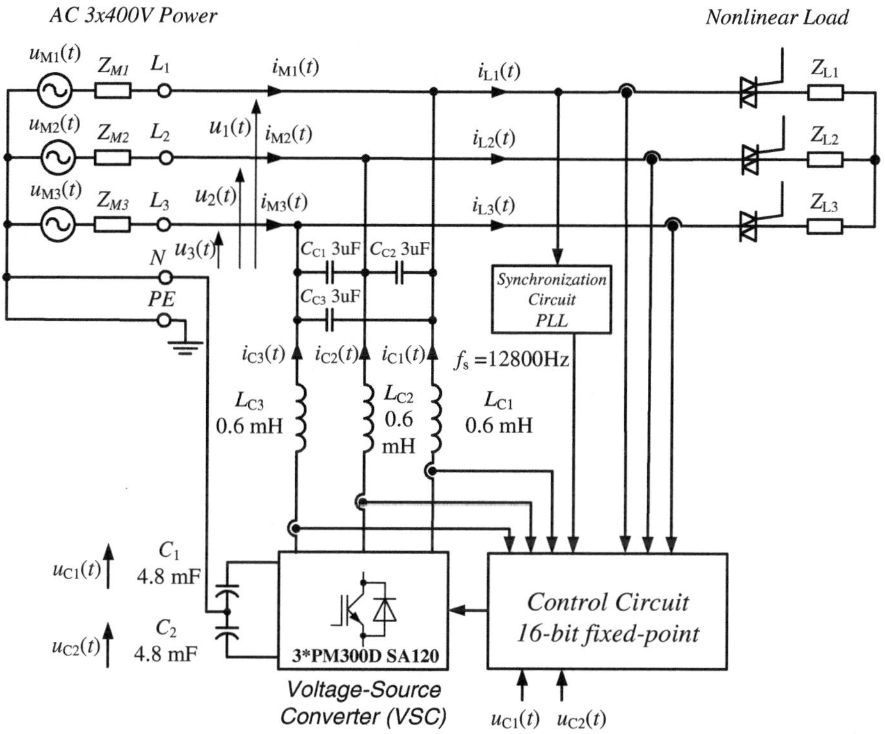

Fig. 4.21 Three-phase compensation circuit with shunt APF (with unity gain)

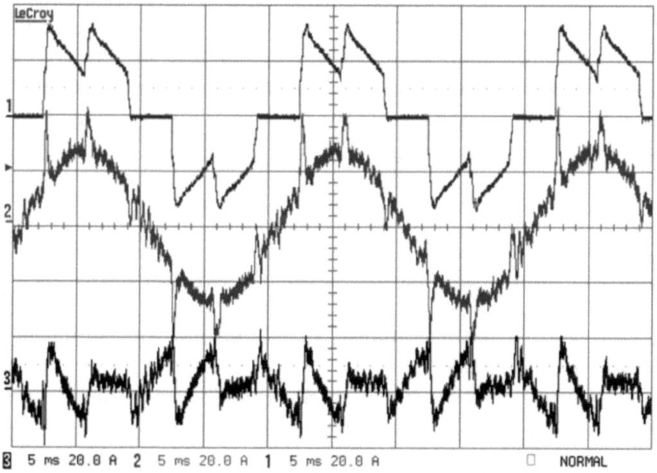

Fig. 4.22 Experimental waveforms (for one phase) of the compensation circuit with the APF in steady-state with the resistive load, from the *top* to the *bottom*: load current i_L, compensating current i_C, line current i_M

design of the control circuit [29]. The APF consists of signal processing control and output circuit with voltage-source converter (VSC). The APF control circuit should force the VSC to behave as a controlled current source. The output circuit consists of two kinds of energy storage components: inductors L_{C1}, L_{C2}, L_{C3} and two DC capacitors C_1, C_2.

The active power filter injects the harmonic currents $i_{C1}(t)$, $i_{C2}(t)$, $i_{C3}(t)$ into the power network and offers a notable compensation for harmonics, reactive power and unbalance. The filter is designed for three or four wire loads. The nonlinear load consists of a thyristor power controller with a resistive load. Experimental waveforms of the compensation circuit in steady-state are shown in Fig. 4.22.

The control algorithm for the proposed APF is based on the strategy resulting from the instantaneous reactive power theory initially developed by Akagi et al. [30–32]. A simplified block diagram for the APF control algorithm is depicted in Fig. 4.23, based on the circuit designed in Sozański et al. [29]. The algorithm is realized using the digital signal processor TMS320C50 which has sampling rate f_s. The digital signal processor is synchronized using the PLL circuit (Fig. 4.14) with the line voltage U_1 and the algorithm is performed N_M times per line period. The sampling periods can be calculated with the formula:

$$T_s = \frac{T_M}{N_M} \tag{4.28}$$

where: T_M—period of the line voltage, $f_M = T_M^{-1}$—frequency of the line voltage, N_M—total number of samples per line voltage period.

For the line voltage frequency of $f_M = 50$ Hz and the number of samples chosen to $N_M = 256$, the sampling period is equal to $T_s = 78.125$ μs and the sampling rate is equal to $f_s = 12{,}800$ samples/s.

Three-phase current signals can be transformed into the equivalent two-phase representation. The transformation (**1-2-3 → α-β -0**) from the three-phase current signals $i_{L1}(nT_s)$, $i_{L2}(nT_s)$ $i_{L3}(nT_s)$, to the two-phase $i_{L\alpha}$, $i_{L\beta}$ with an additional neutral signal i_{L0} can be written into a matrix form as:

$$\begin{bmatrix} i_{L\alpha}(nT_s) \\ i_{L\beta}(nT_s) \\ i_{L0}(nT_s) \end{bmatrix} = \sqrt{\frac{2}{3}} \begin{bmatrix} 1 & -1/2 & -1/2 \\ 0 & \sqrt{3}/2 & -\sqrt{3}/2 \\ 1/\sqrt{6} & 1/\sqrt{6} & 1/\sqrt{6} \end{bmatrix} \begin{bmatrix} i_{L1}(nT_s) \\ i_{L2}(nT_s) \\ i_{L3}(nT_s) \end{bmatrix} \tag{4.29}$$

where: $i_{L\alpha}(nT_s)$—digital representation of signal $i_{L\alpha}$ for sampling period T_s, n—index of the current sample.

In the next step, the two-phase signals are transformed from the rotating to the stationary reference frame. This transformation is commonly called the reverse Park transformation and can be digitally calculated by equations:

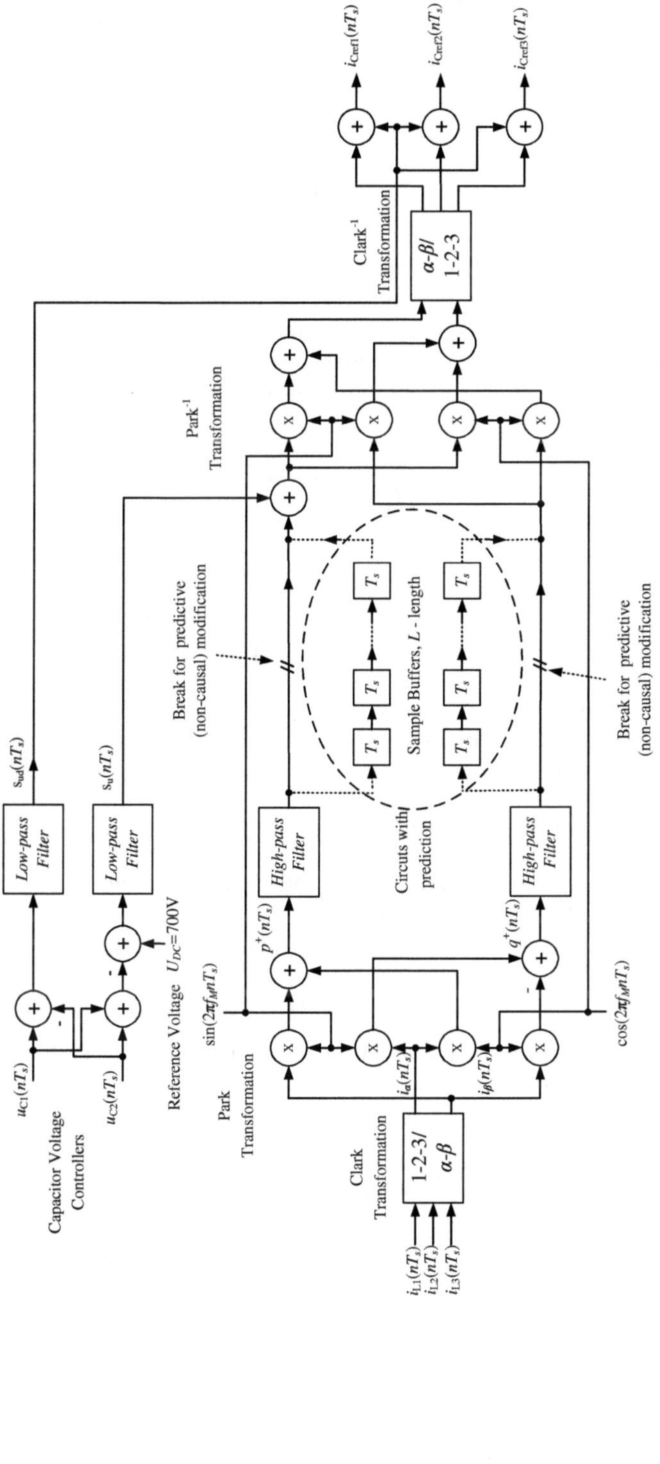

Fig. 4.23 Block diagram of control algorithm

$$\begin{cases} p^{+}(nT_s) = i_{L\alpha}(nT_s)\sin\left(\dfrac{2\pi n}{N_M}\right) - i_{L\beta}(nT_s)\cos\left(\dfrac{2\pi n}{N_M}\right) \\ q^{+}(nT_s) = i_{L\alpha}(nT_s)\cos\left(\dfrac{2\pi n}{N_M}\right) + i_{L\beta}(nT_s)\sin\left(\dfrac{2\pi n}{N_M}\right) \end{cases} \tag{4.30}$$

and the digital sinusoidal reference signal is given by the formula:

$$\sin\left(2\pi f_M nT_s\right) = \sin\left(2\pi f_M n\frac{T_M}{N_M}\right) = \sin\left(\frac{2\pi n}{N_M}\right) \tag{4.31}$$

In order to generate the reference sinusoidal and cosinusoidal signals, a table containing sinus function values is allocated in the digital signal processor program memory. Signal $p^{+}(nT_s)$ represents instantaneous active power and signal $q^{+}(nT_s)$ represents instantaneous reactive power. The DC components of signals $p^{+}(nT_s)$ and $q^{+}(nT_s)$ are removed by a high-pass digital IIR filter. The high pass filter is described by equations:

$$H(z) = \frac{b - bz^{-1}}{1 + az^{-1}}$$

and

$$b = \frac{2(T_1/T_s)}{1 + 2(T_1/T_s)},\ a = \frac{1 - 2(T_1/T_s)}{1 + 2(T_1/T_s)} \tag{4.32}$$

where: $T_1 = 0.016$ s—the reference (analog) filter time constant.

For stabilizing the DC voltage a proportional controller is used, and its response is calculated by the equation:

$$s_U(nT_{p1}) = k_P\left(U_{DC} - \left(u_{C1}(nT_{p1}) + u_{C2}(nT_{p1})\right)\right) \tag{4.33}$$

where: u_{C1}, u_{C2}—voltage on capacitor C_1 and C_2, respectively, k_P—gain of voltage controller, U_{DC}—DC reference voltage.

Signal $s_U(nT_s)$ is subtracted from the component $p^{+}(nT_s)$:

$$p_C^{+}(nT_s) = p^{+}(nT_s) - s_U(nT_s) \tag{4.34}$$

In the next step components p_C^{+} and q^{+} are transformed by Park transformation into the two-phase representation:

$$\begin{cases} i_{Cref\alpha}(nT_s) = p_C^{+}(nT_s)\sin\left(\dfrac{2}{N_M}\right) + q^{+}(nT_s)\cos\left(\dfrac{2\pi n}{N_M}\right) \\ i_{Cref\beta}(nT_s) = -p_C^{+}(nT_s)\cos\left(\dfrac{2\pi n}{N_M}\right) + q^{+}(nT_s)\sin\left(\dfrac{2}{N_M}\right) \end{cases}]]> \tag{4.35}$$

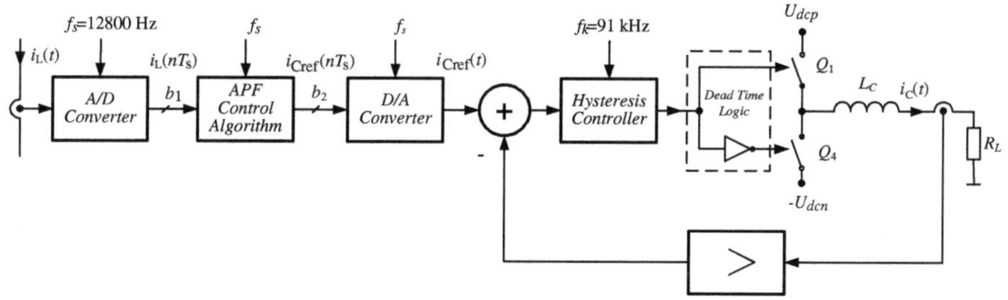

Fig. 4.24 Block diagram of control circuit for one phase

Fig. 4.25 Block diagram of hysteresis current controller algorithm with variable width of the hysteresis

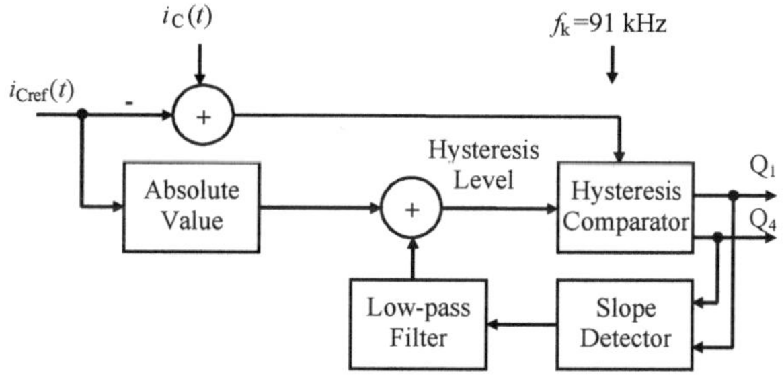

and then transformed back to the three-phase reference current signals:

$$\begin{bmatrix} i_{Cref1}(nT_s) \\ i_{Cref2}(nT_s) \\ i_{Cref3}(nT_s) \end{bmatrix} = \sqrt{\frac{3}{2}} \begin{bmatrix} 2/3 & 0 & \sqrt{6}/3 \\ -1/3 & \sqrt{3}/3 & \sqrt{6}/3 \\ -1/3 & -\sqrt{3}/3 & \sqrt{6}/3 \end{bmatrix} \begin{bmatrix} i_{Cref\alpha}(nT_s) \\ i_{Cref\beta}(nT_s) \\ i_{L0}(nT_s) \end{bmatrix} \tag{4.36}$$

In the next step the output compensation reference current signals $i_{\mathrm{Cref1}}(nT_s)$, $i_{\mathrm{Cref2}}(nT_s)$ $i_{\mathrm{Cref3}}(nT_s)$ are converted to analog form by D/A converter. The control circuit block diagram is shown in Fig. 4.24.

Finally, the output compensation reference current signals are transformed to transistor controlling pulses by current controller. Initially, in the proposed circuit a hysteresis current controller algorithm realized using analog comparators and additional glue logic is employed. The hysteresis control algorithm is based on a nonlinear feedback loop with two-level hysteresis comparators. The inverter switching speed depends largely on the load parameters. In the proposed APF advanced hysteresis current controller with variable width of the hysteresis (Fig. 4.25) is applied. It has additional improvements:

- maximum switching speed is limited;
- switching speed depends on the 'speed history';
- switching speed is dependent on the compensation reference current signals i_{CR1}; the higher the signal level, the lower is the switching speed

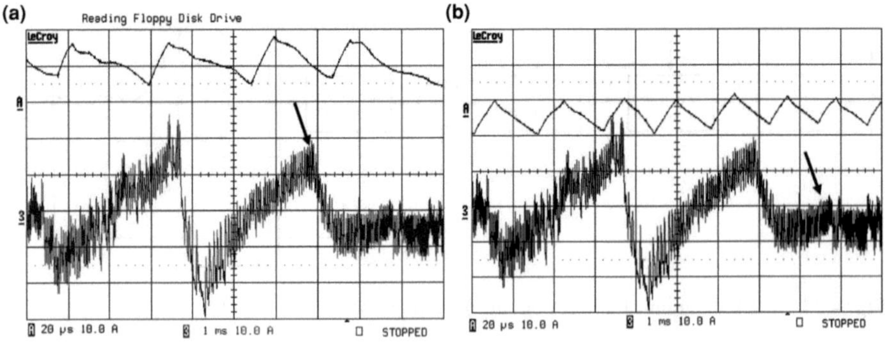

Fig. 4.26 Experimental waveforms, illustration of the dependence of compensation reference current signals $i_{Cref}(t)$ level on the switching speed: **a** for high level of signal, **b** for low level of signal

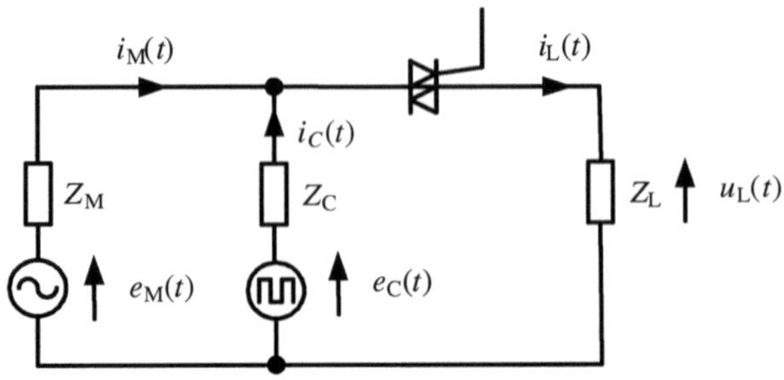

Fig. 4.27 Simplified diagram of a one phase compensation circuit with APF

Figure 4.26 shows dependence of compensation reference current signals $i_{Cref}(t)$ level on the switching speed. For lower signal level the switching speed is around 25 kHz, and for higher level of signal is around 10 kHz.

4.4.2 Predictive Control Algorithm for APF

Experimental waveforms of the compensation circuit using shunt APF in steady-state are shown in Fig. 4.22. When the value of load current changes rapidly, as in current i_L in Fig. 4.22, the APF transient response is too slow and the line current i_M suffers from dynamic distortion. This distortion causes an increase in harmonic content in the line current, which is dependent on a time constant. In the circuit shown in Fig. 4.21 the total harmonic distortion ratio (THD) is increased by more than 10%. The APF control current dynamics is mainly dependent on the inverter output time constant T_o, consisting of APF output impedance and resultant impedance of load and line. Problems with the APF dynamics are not very widely described in the literature, but it is possible to find a few publications: e.g.,

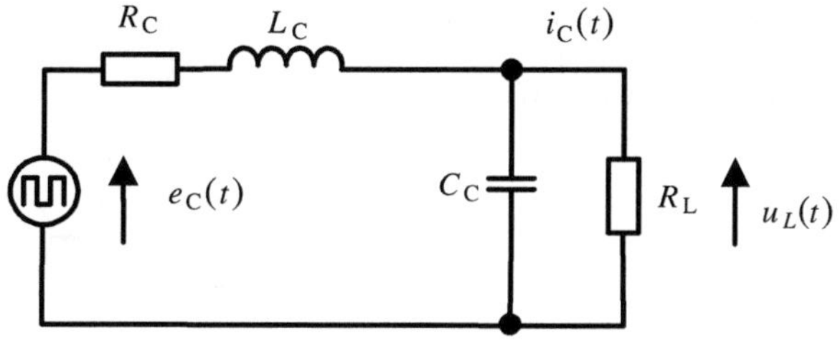

Fig. 4.28 Simplified diagram of the APF output circuit

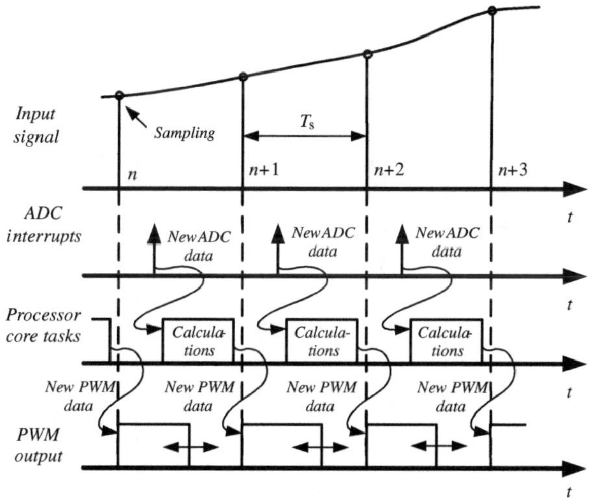

Fig. 4.29 Timing diagram for digital realization of APF control circuit

Mariethoz and Rufer [33], Marks and Green [34], Wojciechowski and Strzelecki [35], Sozański [36–41].

The primary problem is to develop a predictive control algorithm which allows a reduction in the impact of dynamic distortion and ultimately reduces the line current THD. The power loads can be divided into two main categories: predictable loads and noise-like loads. Most loads belong to the first category. For this reason it is possible to predict current values in subsequent periods, after a few periods of observation.

4.4.2.1 APF Output Circuit

The compensation circuit shown in Fig. 4.21 can be simplified to the circuit depicted in Fig. 4.27 (for one phase). Z_M represents resultant line impedance, Z_C represents APF output impedance and Z_L represents load impedance. The circuit time constant is mainly dependent on Z_C, therefore it is possible to simplify the circuit to the one presented in Fig. 4.28. In the author's opinion this simplified circuit allows the simulation of a system dynamic with sufficient accuracy.

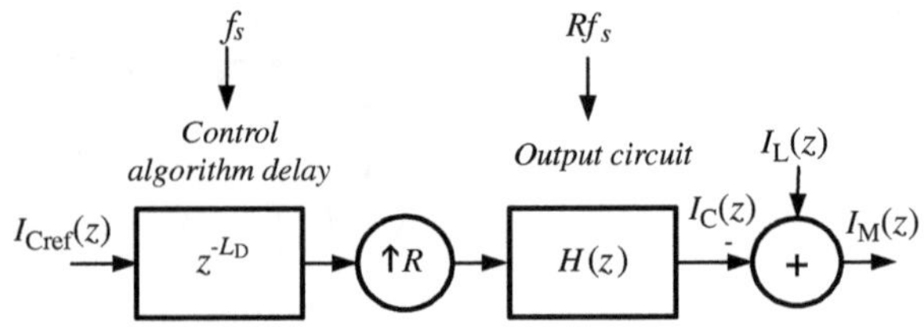

Fig. 4.30 Block diagram for digital simulation of compensation circuit (for one phase)

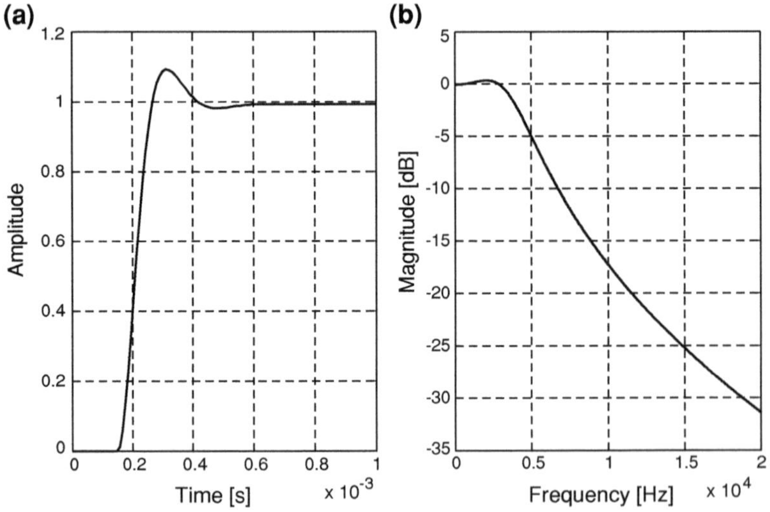

Fig. 4.31 The APF output circuit: **a** step response, **b** frequency response

Voltage transmittance of such a circuit is described by the formula:

$$H(s) = \frac{U_L(s)}{U_C(s)} = \frac{R_L}{R_C + R_L + (R_C R_L C_C + L_C)s + R_L L_C C_C s^2} \tag{4.37}$$

The whole APF control circuit is realized using a 16-bit fixed-point DSP, therefore it is sensible to realize a digital simulation of the entire compensation system. The voltage transmittance of the analog circuit described by Eq. 4.36 is converted into digital domain using bilinear transformation. Due to the greater accuracy of simulation, the response of the output circuit is calculated R times more often than the whole algorithm, so its sampling frequency is equal to the Rf_s. Transmittance of the digital representation of the APF output circuit for the assumed values of the circuit elements—$R_C = 0.1\ \Omega$, $L_C = 0.6$ mH, $C_C = 3\ \mu$F, $R_L = 12\ \Omega$, oversampling ratio $R = 8$—and the sampling frequency $Rf_s = 8 \times 12{,}800$ Hz, is described by the equation:

$$H(z) = \frac{U_L(z)}{U_C(z)} = \frac{0.01152 + 0.02304z^{-1} + 0.01152z^{-2}}{1 - 1.716z^{-1} + 0.7627z^{-2}} \tag{4.38}$$

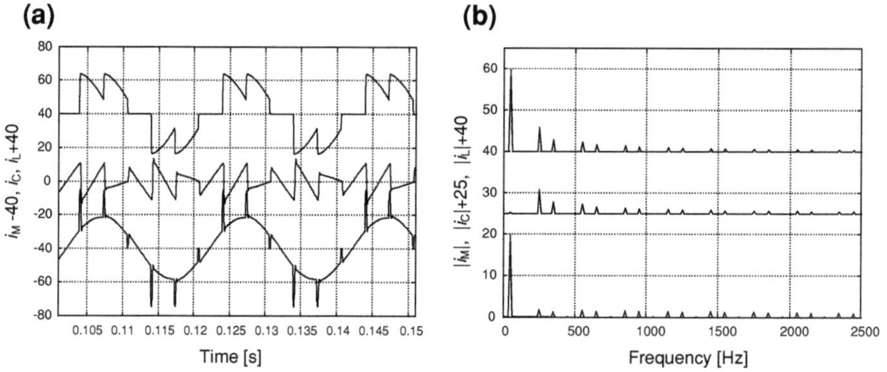

Fig. 4.32 Simulation currents of compensation circuit, load current i_L (*top*), compensating current i_C (*middle*), line current i_M (*bottom*): **a** waveforms, **b** spectrum

Table 4.3 Line current parameters

Current i_M	$I_{M(rms)}$ (A)	*THD* (%)	*SINAD* (dB)	THD_{50} (%)
Without compensation	15.03	39.33	−8.73	38.98
Classical algorithm	14.53	27.75	−11.46	27.24
Prediction algorithm, $T_{AH} = 214$ µs	14.00	3.24	−29.78	1.32

A timing diagram of the digital realization of the APF control circuit is shown in Fig. 4.29. Analog signal is sampled at a frequency f_s, and then is converted to digital form by an analog–digital converter. This digital signal is later processed by the main control algorithm APF and then it is sent to the PWM modulator. The minimum delay of the entire control algorithm is, therefore, at least one sampling period T_s. This delay must be taken into account during the simulation. For the case under consideration the algorithm delay time equal to two sample periods was chosen, $L_D = 2$. The block diagram for digital simulation of the compensation circuit is shown in Fig. 4.30. The step response of such a circuit is depicted in Fig. 4.31a (for $I_L(z) = 0$). The total delay time of this circuit is equal to about 0.25 ms. The frequency response is shown in Fig. 4.31, the cut-off frequency value of this circuit is equal to about 3.5 kHz. For the simulation, the reference compensating signal $I_C(z)$ is calculated by an instantaneous power algorithm, and load current $I_L(z)$ is calculated by the power controller with resistive load simulator.

The simulated line current $I_M(z)$ is the result of adding both signals. Simulated current signal waveforms of the compensating circuit are shown in Fig. 4.32. The waveforms obtained from the simulation are very close to the experimental waveforms (Fig. 4.22). For simplicity, in the simulation model, the PWM modulator is omitted, so the waveforms do not have modulation components. The spectrums of currents the i_L, i_C, i_M are shown in Fig. 4.32b. In Table 4.3, there are presented some line current i_M parameters. For example the THD of simulated current is equal to 39.33% for a circuit without compensation and is equal up to

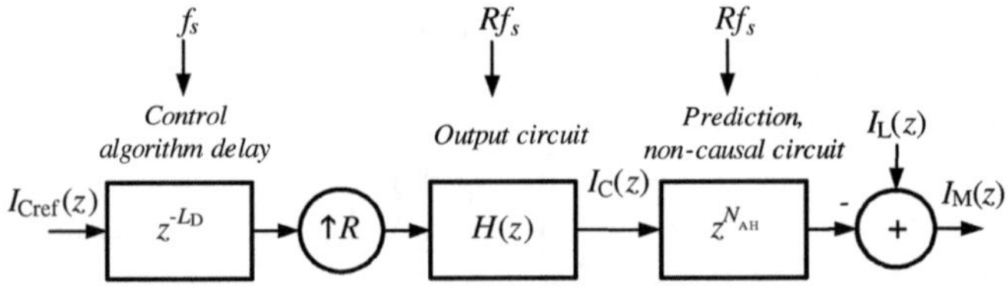

Fig. 4.33 Block diagram of compensation circuit digital simulation with non-causal current predictor (for one phase)

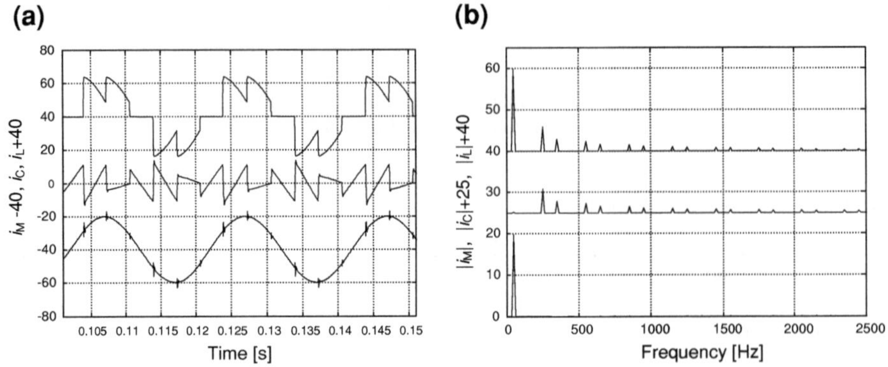

Fig. 4.34 Simulation currents of compensation circuit with prediction, load current i_L (*top*), compensating current i_C (*middle*), line current i_M (*bottom*): **a** waveforms, **b** spectrum

27.75% for a circuit with compensation. This shows the compensation circuit is not working effectively with the rapid changes in load current. In order to, reduce the impact of this phenomenon in a practical compensation circuit, the rate of change of load current is reduced by adding a serial inductor. However, this solution increases the weight and cost of the compensation system.

4.4.2.2 Solution

For predictable loads the compensation error depends on the dynamics of the output circuit and occurs periodically, so it can be partially compensated by sending a compensating current in advance. Such a solution would be impossible to implement in an analog control system, but it is easy to implement in a digital control system. In the proposed solution, for predictable loads, it is possible to use a circuit with prediction [36–41], as shown in Fig. 4.33. The desired digital compensation reference current signal $I_{Cref}(z)$ is sent to the circuit using prediction. Previous current compensation signal samples are stored in memory, and sent to present output in advance. This compensation is dependent on the inverter output time constant. Because the time constant is mainly dependent on the output

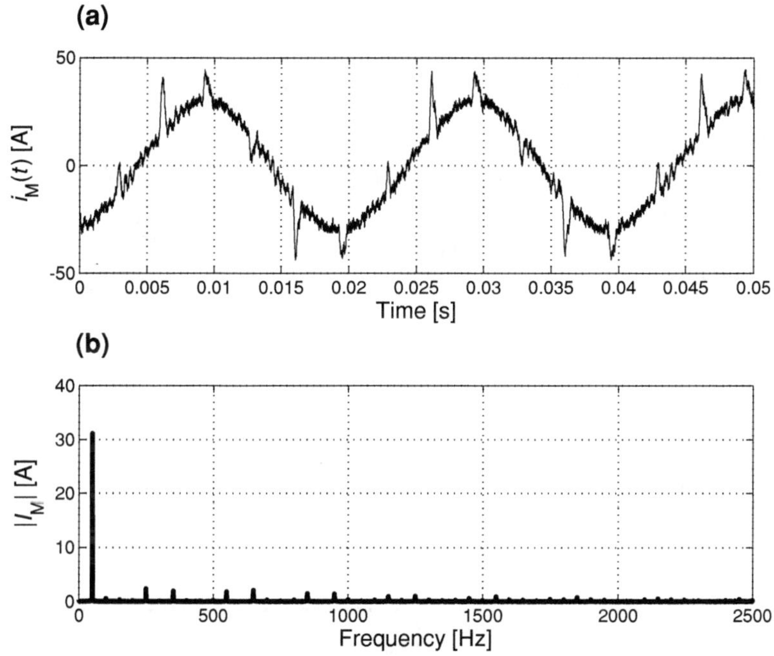

Fig. 4.35 Experimental waveforms of the classical three-phase APF in steady-state with the resistive load line current $i_{M1:}$ **a** waveform, **b** normalized spectrum magnitude

inverter inductor value, it is possible to set a constant value for advance time T_{AH}. In the considered APF the discrete advance time T_{AH} is:

$$T_{AH} = N_{AH} T_s \tag{4.39}$$

where: N_{AH}—number of samples sent ahead.

In the simulated compensation circuit the digital advance time is equal to $T_{AH} = 214$ μs. Simulated current signal waveforms for the compensation circuit with prediction are shown in Fig. 4.34a. The waveforms of line current i_M show that the goal of compensation has been reached and that the shape of the current is very close to sinusoidal. This is also evidenced by the spectrum shown in Fig. 4.34b. Similarly, the same is true with the signal parameters listed in Table 4.3. As a result of the prediction algorithm, the THD was reduced from 27.75 to 3.24%.

4.4.2.3 Experimental Results

For experimental studies the compensation system shown in Fig. 4.21 was used. A simplified block diagram for the active power filter control algorithm is depicted

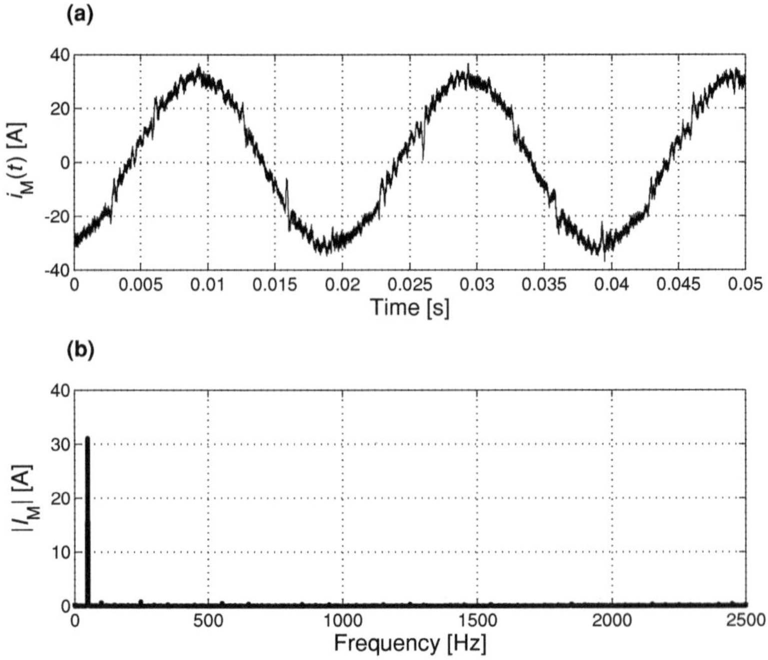

Fig. 4.36 Experimental waveforms of the modified three-phase APF in steady-state with the resistive load, line current i_{M1} for $N_{AH} = 3$: **a** waveform, **b** normalized spectrum magnitude

in Fig. 4.23. The implementation of the control algorithm using a fixed-point 16-bit digital signal processor is described in detail in Sect. 4.4.1. Modification of the instantaneous reactive power control algorithm is shown in Fig. 4.23, with two prediction (non-causal) circuits being added to the classical control circuit. Figures 4.35 and 4.36 show the experimental waveforms in the same steady-state conditions, for the classical APF (Fig. 4.35a), and for the modified APF circuit with predictive current compensation (Fig. 4.36a). In these figures are depicted the line current i_{M1} waveform and its normalized spectrum magnitude.

The best results are achieved for the number of samples sent ahead equaling three $N_{AH} = 3$; for a higher number of samples the compensator is unstable. Using the new control algorithm with predictive harmonic compensation it is possible to decrease the harmonic contents in power line currents from around THD $= 22\%$ to near THD $= 5\%$ (for $N_{AH} = 3$).

4.4.2.4 Step Response of APF

The first question in using the APF with a predictive circuit is how it will work when the value of load current changes rapidly. For this reason a step response of load currents for the considered shunt APF is investigated. Figure 4.37a shows the

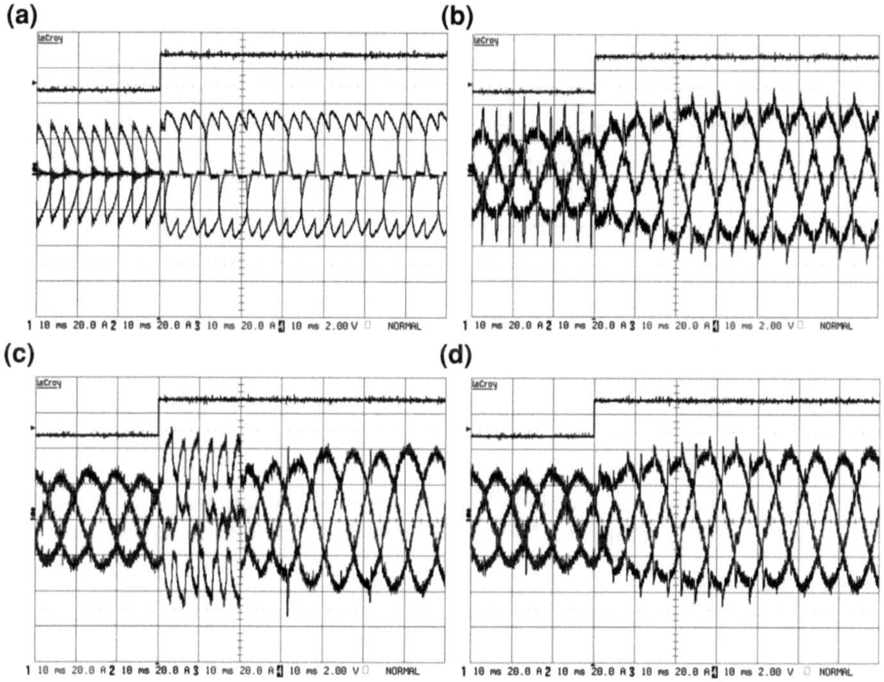

Fig. 4.37 Experimental waveforms of line currents: i_{M1}, i_{M2}, i_{M3} for step response of the load currents for the power controller with the resistive load: **a** APF is switched off, **b** APF with classical control algorithm, **c** APF with predictive algorithm constantly switched on, $N_{AH} = 3$, **d** APF with adaptive predictive algorithm, $N_{AH} = 3$

waveforms of a load current regulated by a power controller with resistive load without the APF compensation. The load current value is adjusted by a power controller regulated by thyristor phase control, by changing trigger angle. The angle is controlled by a square signal (curve 4) shown above in Fig. 4.37. The step response of the load currents for the considered shunt APF with classical control algorithm is shown in Fig. 4.37b, for the same load current values as in Fig. 4.37a. The APF step response setting time is equal to one period of line voltage. Figure 4.37c shows step response of the APF with modified control algorithm with predictive circuits constantly switched on. In the first line voltage period, after the value of load currents is changed, the APF sends compensating current i_c adequate to previous load currents and the resultant currents are non-sinusoidal (Fig. 4.37c). Therefore the predictive circuit should be modified to the circuit presented in Fig. 4.38. In this circuit current samples $I_{in}(z)$ are stored in DSP memory, in N_M—sample buffer and the next period of line current is compared with present samples, then if the absolute difference of the present sample value and respective sample stored in memory is less than an assumed value, the predictive (non-causal) current compensation algorithm is switched on

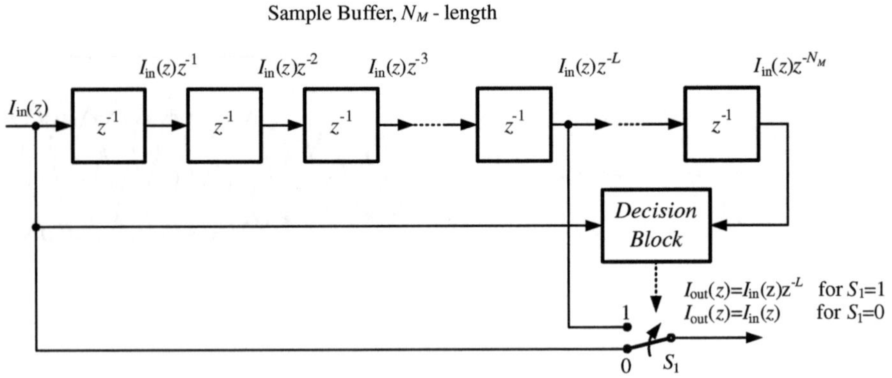

Fig. 4.38 Block diagram of adaptive circuit with prediction

(switch S_1 in position 1). To the output are sent non-causal samples $I_{in}(z^{-L})$ from the buffer. The length of sample buffer can be calculated by the formula:

$$L = N_M - N_{AH} \tag{4.40}$$

If load current is changed, the non-causal current predictive algorithm is switched off (switch S_1 in position 0), the algorithm waits for a steady-state and once detected, it switches on again (switch S_1 in position 1). The output signal is described by the formula:

$$\begin{cases} I_{out}(z) = I_{in}(z^{-L}) & \text{for } S_1 = 1 \\ I_{out}(z) = I_{in}(z) & \text{for } S_1 = 0 \end{cases} \tag{4.41}$$

Waveforms with result of such modification are shown in Fig. 4.37d. After the rapid change in value of load current, the predictive (non-causal) circuits are switched off. In the considered case (Fig. 4.37d), it is again switched on after two and a half line voltage periods.

Using adaptive predictive current compensation it is possible to decrease harmonic content for predictable loads. This modification for an instantaneous reactive power control algorithm is very simple and additional computational workload is very small. Therefore it is easy to implement it in an existing control circuit based on digital signal processor, microcontroller or programmable digital circuit (FPGA, CPLD, etc.). This simple modification can be easily implemented in existing APF digital control circuits, improving the quality of harmonic compensation. The considered current prediction circuit may also be useful for other APF control algorithms. What is more, the presented circuits and methods can be useful for other power electronics applications, such as serial APF, power conditioners, high quality AC sources, etc.

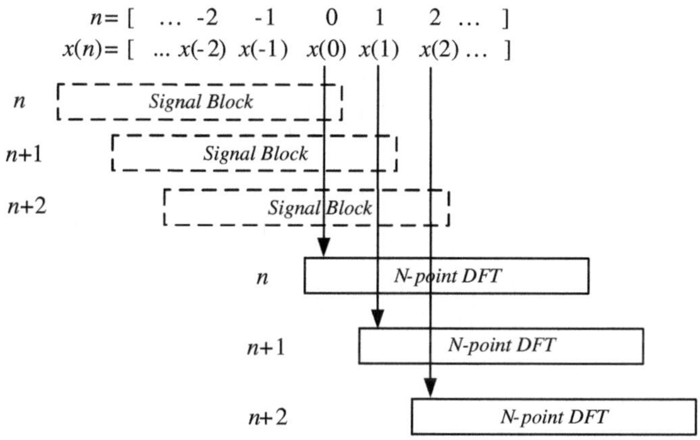

Fig. 4.39 Block diagram of signal flow for typical DFT

Fig. 4.40 Block diagram of single-bin sliding DFT filter

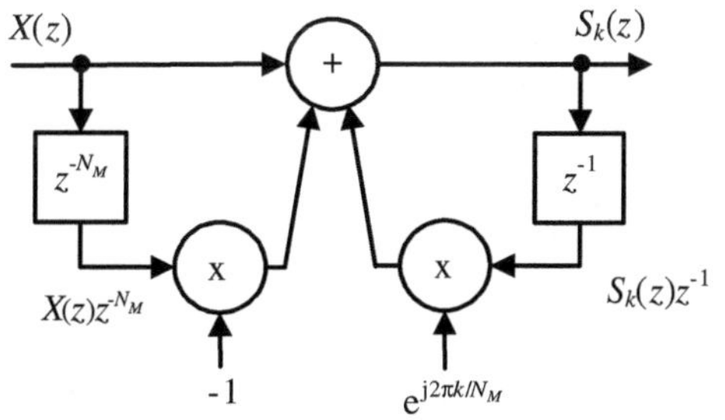

4.4.3 Digital Control Algorithm Based on Sliding DFT

The active power filter injects the harmonic current i_C, into the power network and offers a notable compensation for harmonics and reactive power. The compensating current can be determined by:

$$i_C(t) = i_L(t) - I_{m1} \sin(2\pi f_M t + \varphi_1) \tag{4.42}$$

where: I_{m1}—amplitude of first harmonic.

In the case when harmonics compensation is perfect the line current $i_M(t)$ consists of the first harmonic line current only:

$$i_M(t) = I_{m1} \sin(\omega t + \varphi_1) \tag{4.43}$$

When phase angle between line voltage u_1 and line current i_S is equal to zero the reactive power is compensated too. Spectrum analysis of signal in power electronics is an important measuring technique. The usual method for spectrum analysis is the DFT and its efficient implementation, the fast Fourier transform

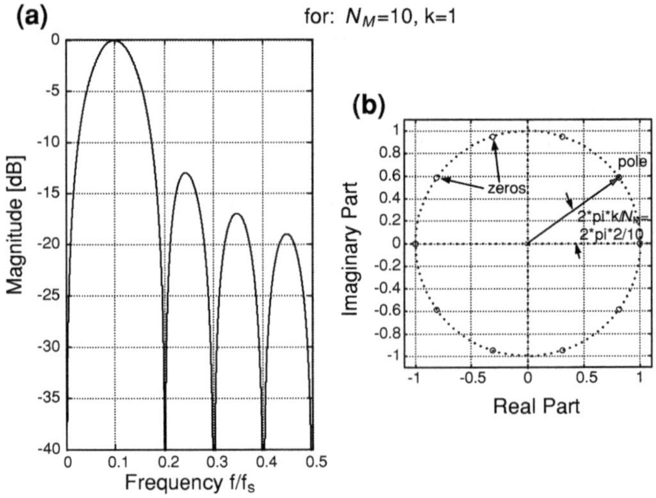

Fig. 4.41 Sliding DFT characteristics for $N_{\mathrm{M}} = 10$ and $k = 1$: **a** magnitude, **b** z-domain pole/zero location

Fig. 4.42 Block diagram of guaranteed-stable sliding DFT filter

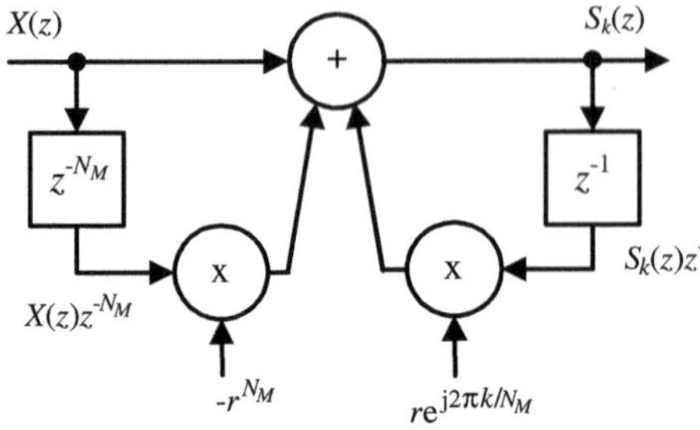

(FFT). Typically, for every sample of input signal N_{M}—point, DFT is calculated as shown in Fig. 4.39.

For this kind of task, the sliding DFT algorithm is better than ordinary DFT. It is very simple and efficient, especially in a coherent sampling case. The sliding DFT algorithm is well described by Jacobsen and Lyons [42]. The z-domain transfer function for the kth bin of the sliding DFT filter is described by the equation:

$$H_{\mathrm{SDFT}}(z) = \frac{1 - z^{-N_{\mathrm{M}}}}{1 - e^{j2\pi k/N_{\mathrm{M}}} z^{-1}} \tag{4.44}$$

where: N_{M}—length of the signal block, typically equal to signal period, k—number of frequency bin (Fig. 4.40).

The magnitude frequency characteristic of single-bin sliding filter for $N = 10$ and $k = 1$ is shown in Fig. 4.41a. The passband and stopband are very poor but

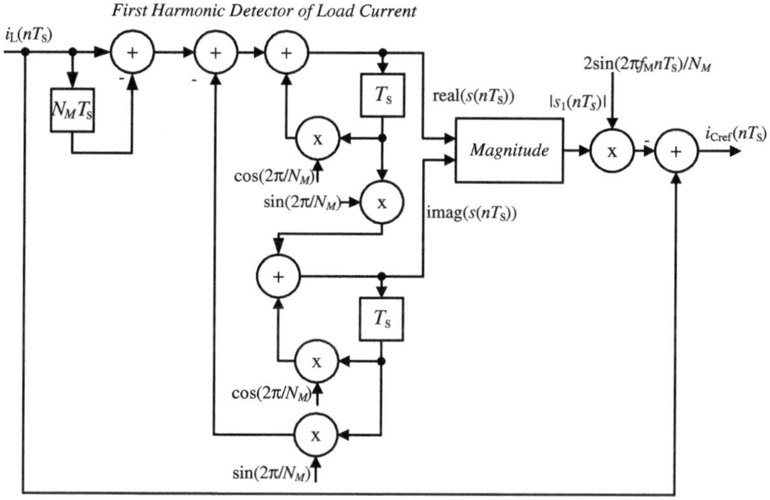

Fig. 4.43 Simplified block diagram for the APF control algorithm using sliding DFT first harmonic detector

they are adequate for coherent sampled signals. The sliding DFT filter is only marginally stable because its pole resides on the z-domain's unit circle as shown in Fig. 4.41b.

By using damping factor r it is possible to force the pole to be at a radius of r inside the unit circle. The transfer function of this solution is described by equation:

$$H_{SDFT}(z) = \frac{1 - r^{N_M} z^{-N_M}}{1 - r \cdot e^{j2\pi k/N_M} z^{-1}} \tag{4.45}$$

Guaranteed-stable sliding DFT filter structure is depicted in Fig. 4.42. It is useful for low resolution fixed-point calculation, for example in fixed-point DSP and field programmable gate array (FPGA) circuits. For floating-point DSP such as SHARC it is possible using circuit from Fig. 4.40.

In the considered solution only one signal-bin sliding DFT filter structure for detecting the first harmonic of load current is used. The first harmonic spectral component signal of load current is calculated thus:

$$s_1(nT_s) = s_1((n-1)T_s) \cdot e^{j2\cdot\pi/N_M} - i_L((n-N_M)T_s) + i_L(nT_s) \tag{4.46}$$

where: $i_L(nT_s)$—discrete signal representing load current, $s_1(nT_s)$—discrete signal representing first harmonic complex spectral component of first phase load current, $s_1((n-1)T_s)$—discrete signal representing previous first harmonic complex spectral component of load current.

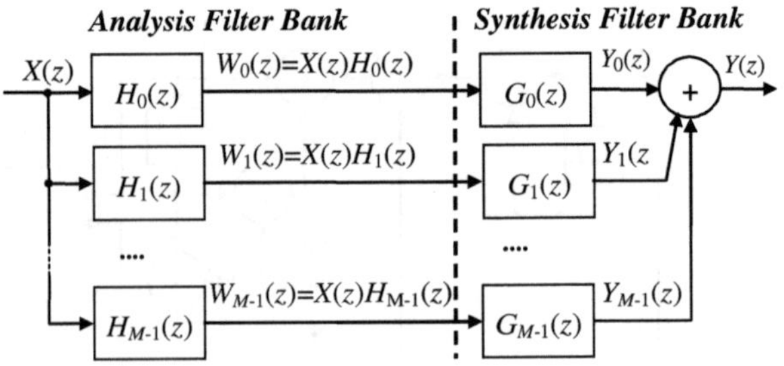

Fig. 4.44 An M-channel band analysis and synthesis filter bank

The discrete signal representing the first harmonic signal of a load current with zero phase angle between line voltage $u_1(t)$ and line current $i_{S1}(t)$ can be described by the equation:

$$i_{H1}(nT_s) = 2/N_M|s_1(nT_s)|\sin(2\pi 50nT_s + \varphi_1) \tag{4.47}$$

Compensating current signal is the result of a difference between the load current signal and the first harmonic reference sinusoidal signal:

$$i_C(nT_s) = i_L(nT_p) - 2/N_M|s_1(nT_s)|\sin(2\pi 50nT_s + \varphi_1) \tag{4.48}$$

The block diagram of this kind of control circuit is shown in Fig. 4.43. Similarly to Sect. 4.2, a prediction algorithm was successfully applied too.

In the case when current imbalance must be compensated, compensating current for the first phase $i_{c1}(nT_s)$ has to be calculated by the formula:

$$i_{C1}(nT_s) = i_{L1}(nT_s) - 2/N\frac{|s_{11}(nT_s)| + |s_{12}(nT_s)| + |s_{13}(nT_s)|}{3}\sin(2\pi 50nT_s + \varphi_1) \tag{4.49}$$

where: $s_{11}(nT_s)$, $s_{12}(nT_s)$, $s_{13}(nT_s)$—discrete signal representing first harmonic complex spectral component signal of first-, second-, and third-phase load current.

In the summing block the resultant magnitude of three-phase current is calculated. The magnitude is used to modulate the amplitude of each phase first harmonic reference signal. By subtracting these signals with appropriate input signals, the output compensating signals $i_{C1}(nT_s)$, $i_{C2}(nT_s)$, $i_{C3}(nT_s)$ are calculated. One of the most difficult tasks for the control algorithm is magnitude calculation using fixed-point arithmetic. Especially the calculation of square root is a potential source of big errors.

In the proposed algorithm the square root is calculated using the following formula:

$$sqrt(x) = -0.2831102x^2 + 1.0063284x + 0.272661 \tag{4.50}$$

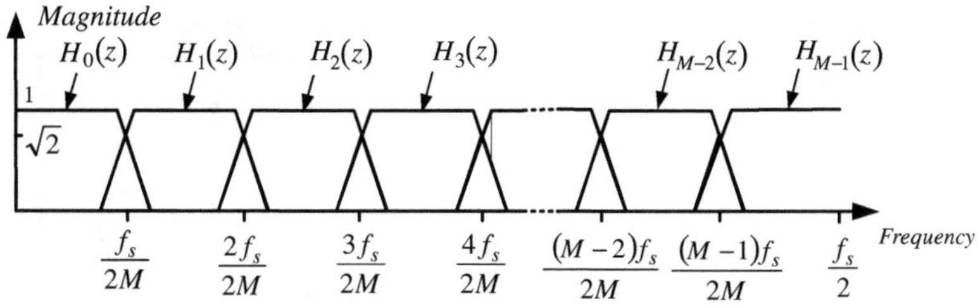

Fig. 4.45 Frequency responses of M-channel uniform band analysis and synthesis filter banks

4.4.4 Algorithm Based on Filter Banks

Analysis filter banks decompose signal spectra into a number of directly adjacent frequency bands and recombine the signal spectra by means of synthesis filter banks. In most cases signals are separated into more than two subband signals. By cascading two-channel filter banks into a tree structure it is possible to obtain a uniform M-channel filter bank with M equaling two. If the two-channel filter bank provides perfect reconstruction then this also holds for the tree structure filter bank. One disadvantage is that the tree structure filter bank possesses high computational complexity. The general form of the M-channel filter banks is shown in Fig. 4.44, where M is the number of subbands. The output signal of filter banks $Y(z)$ can be calculated by the equation:

$$Y(z) = X(z)\sum_{k=0}^{M-1} H_k(z)G_k(z) \tag{4.51}$$

It is possible to simplify this equation to:

$$Y(z) = F(z)X(z) \tag{4.52}$$

where: $F(z)$ denotes the quality of signal reconstruction. If $|F(e^{j\Omega}| = 1$ for all frequencies, the filter bank is without amplitude distortion.

If $F(e^{j\Omega})$ has linear phase (constant group delay), the filter bank is without phase distortion. When $F(z)$ is pure delay, the filter bank is called *perfect reconstruction.* A filter bank with amplitude and/or phase distortion which can be kept arbitrarily small is called a filter bank with *almost perfect reconstruction.* Another function important for the discussed filter banks is power complementary. This ensures representation of whole input signal spectrum in subbands. For an M-channel power complementary filter bank the square sum of transfer functions $H_k(z)$ module is equal to one:

$$\sum_{k=0}^{M-1} |H_k(z)|^2 = 1 \tag{4.53}$$

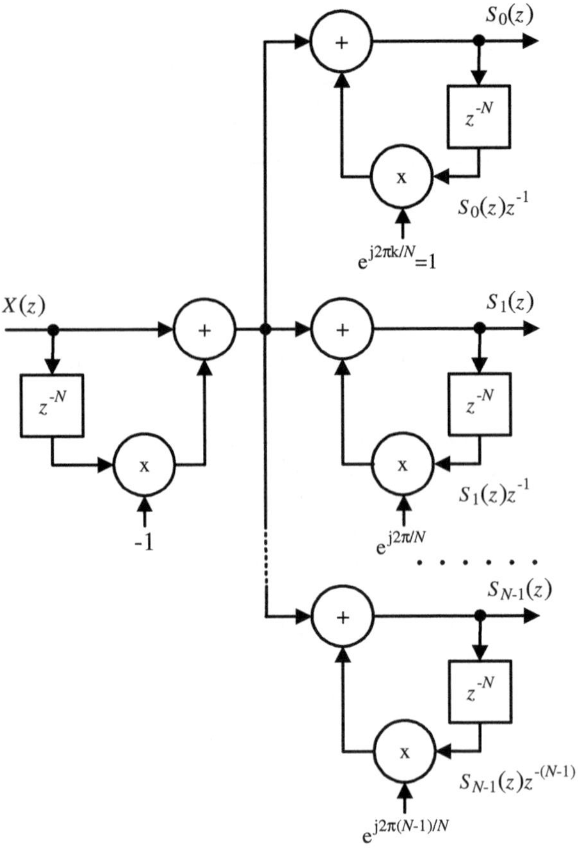

Fig. 4.46 The block diagram of *N*-channel band analysis sliding DFT filter bank

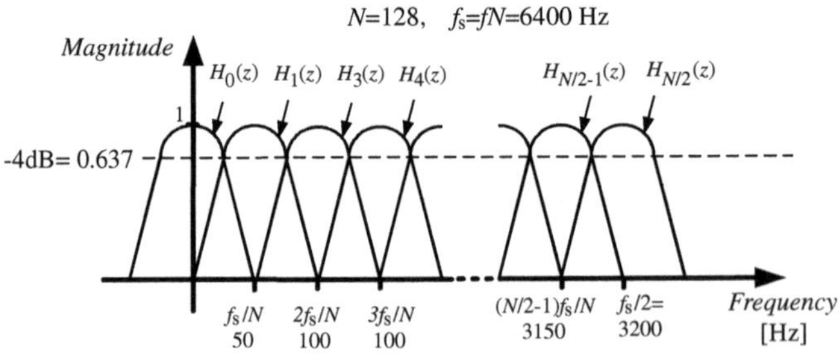

Fig. 4.47 Frequency responses of *N*-channel band analysis sliding DFT filter bank

The typical frequency responses of *M*-channel overlapping uniform band analysis and synthesis filter banks are shown in Fig. 4.45. The theory of filter banks is well described by Vaidyanathan [9], Fliege, [10] and many others.

In the proposed solution the user can select which harmonics are most important in the active power filtration process. This is very important, especially

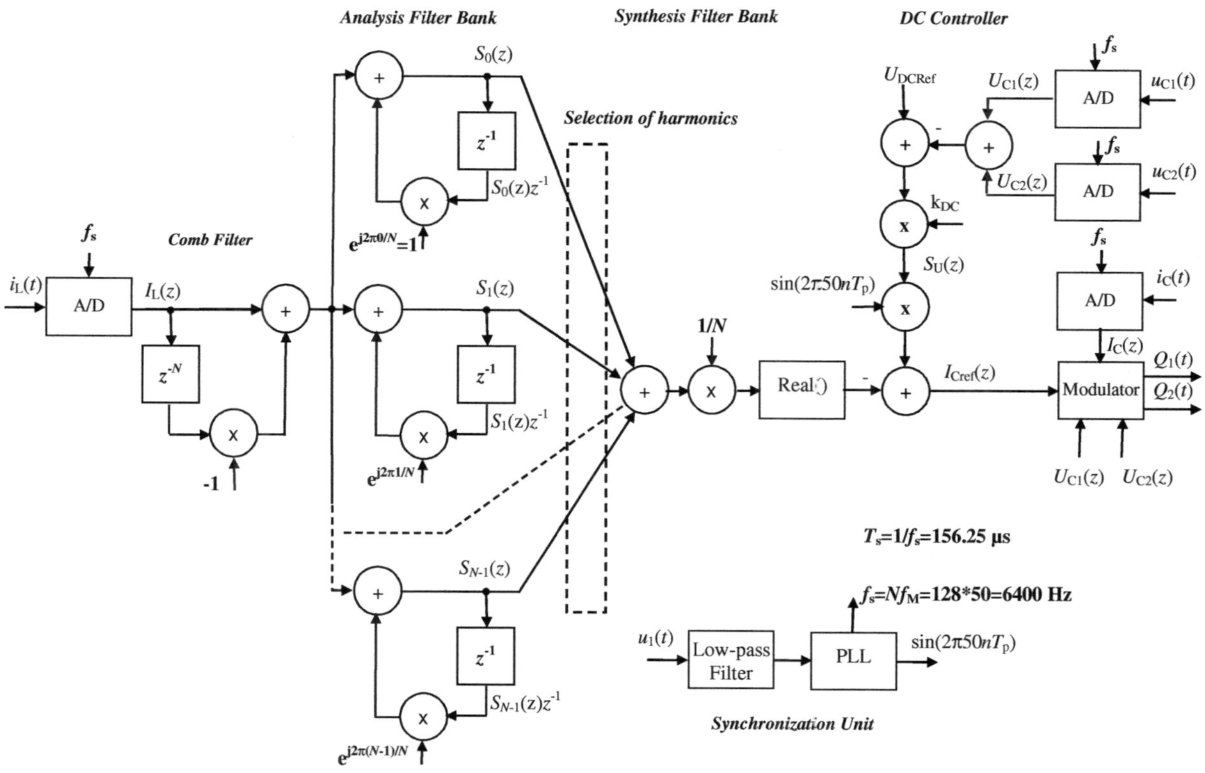

Fig. 4.48 Block diagram APF control circuit using the $N = 128$-channel uniform band analysis and synthesis filter bank using sliding DFT

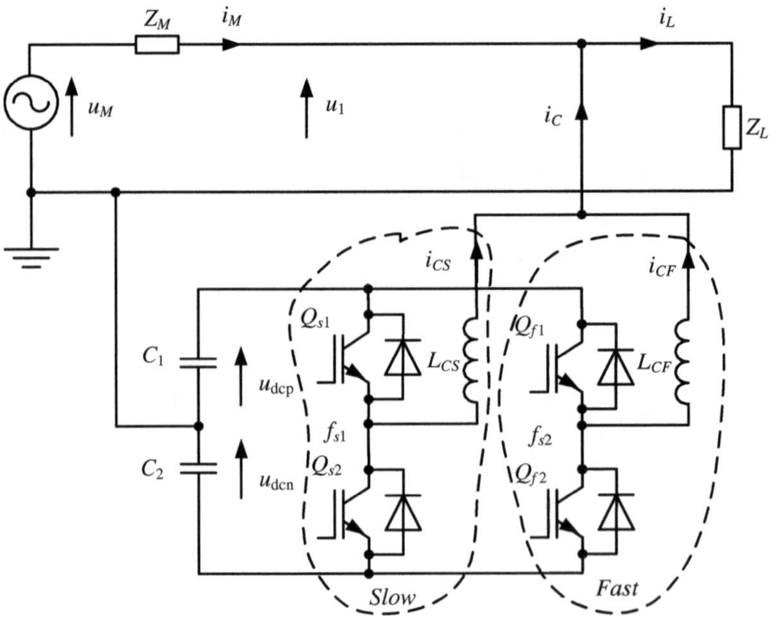

Fig. 4.49 Single-phase active power filter with modified inverter, test circuit

when several APFs are working in parallel or cascaded connection. Sliding DFT algorithm can be used as the analysis filter bank. The block diagram of such a bank is shown in Fig. 4.46. Simplified characteristics of the amplitude of such a filter bank are shown in Fig. 4.47. The APF control circuit using sliding DFT filter banks is shown in Fig. 4.48. The discrete representation of load current signal $i_{\mathrm{L}}(nT_{\mathrm{s}})$ (representation in z-domain $I_{\mathrm{L}}(z)$), is divided into $N = 128$ uniform sub-bands by analysis filter banks. Analysis filter bank consists of common comb filter on the input and the N first order branches. In next stage harmonics are selected for elimination—the set of harmonics (second, third, ...) can be selected and in the synthesis filter banks harmonic signals are synthesised to appropriate current compensating signal.

In the proposed solution, the user can select which harmonics are most important in the active power filtration process. This is very important, especially when several APFs are working in parallel or cascaded connection or for dumping harmonics resonance.

In this case, synthesis filter bank is very simple and consists of only summing block. This kind of solution can be also used for harmonic prediction for canceling permanent dynamics errors (for predictable loads) as is described by Sozański [36].

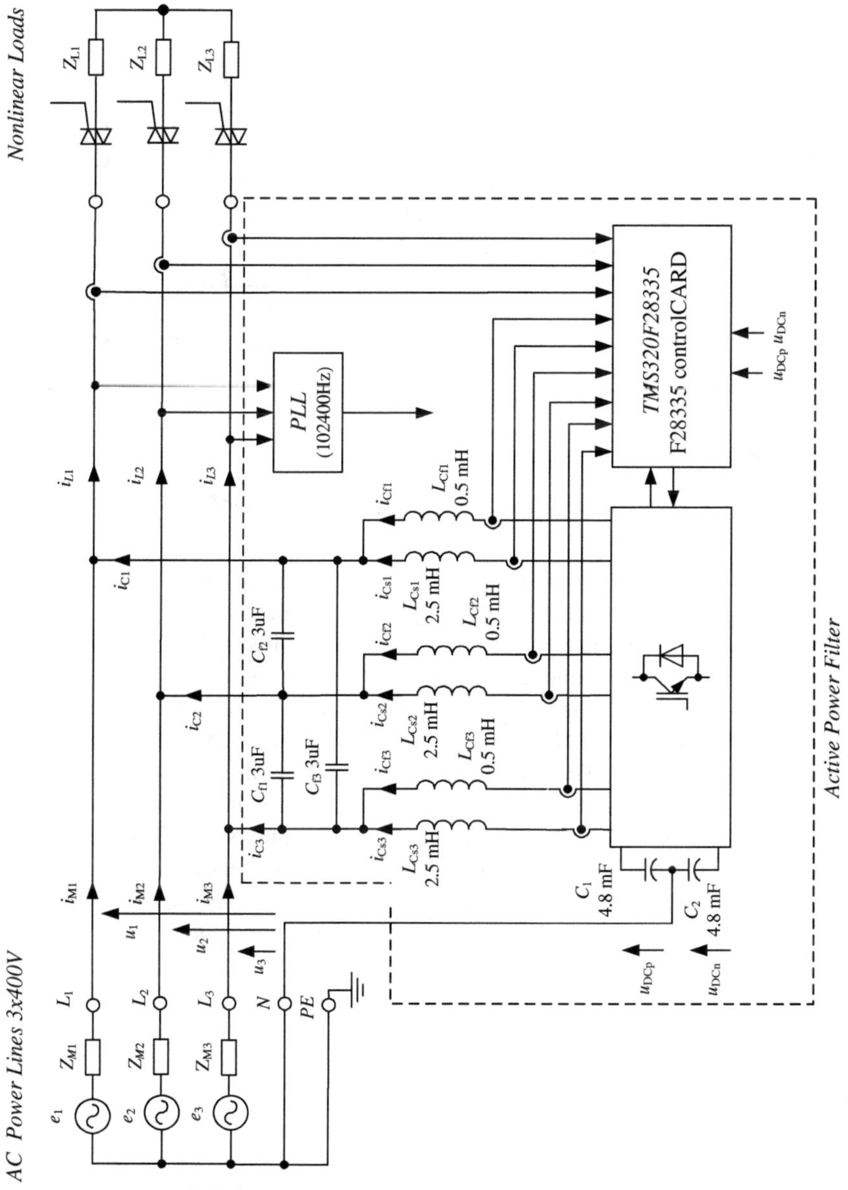

Fig. 4.50 Modified three-phase shunt active power filter

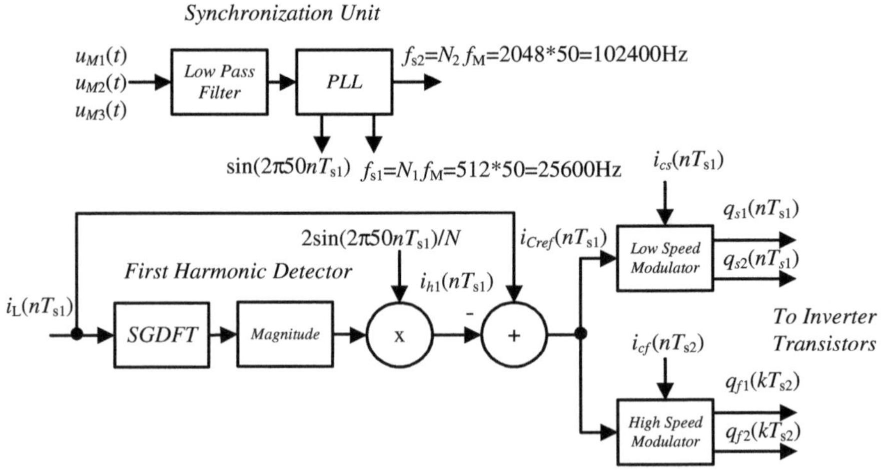

Fig. 4.51 Block diagram of APF control algorithm

4.4.5 Digital Control Algorithm for Multirate APF

Application of prediction algorithm for unpredictable loads is inefficient. Hence the author suggests using multirate APF. Given that high dynamic performance is necessary only for approximately 10% of the time in the line voltage power period, increasing the switching frequency to 60 kHz seems to be unreasonable. Therefore, the author proposes an inverter output stage with two sets of transistors (fast and slow) and inductors [37, 38]. The circuit has a common DC bank (C_1, C_2) for both parts of the inverter. The simplified version of this proposition is shown in Fig. 4.49. The circuit consists of two output stages: one with switches S_{s1}, S_{s2} and inductor L_{Cs}, and a second with switches S_{f1}, S_{f2} and inductor L_{Cf}. The first output stage works continuously with the slowest switching frequency f_{s1}. The value of inductor L_{Cs} is designed to achieve a low i_C current ripple. In the second output stage, switches S_{f1}, S_{f2} work with a several-times higher frequency only in the case when output current changes very quickly (typically 10% of line voltage power period).

4.4.5.1 The Control Circuit

A simplified block diagram of the proposed active power compensation circuit with the parallel APF after improving dynamic performance for a power of 75 kVA is depicted in Fig. 4.50. The circuit consists of a power part with two three-phase IGBT power transistor bridges connected to the *AC* lines through an inductive filtering system composed of inductors L_{Cs1}, L_{Cs2}, L_{Cs3} L_{Cf1}, L_{Cf2}, L_{Cf3}.

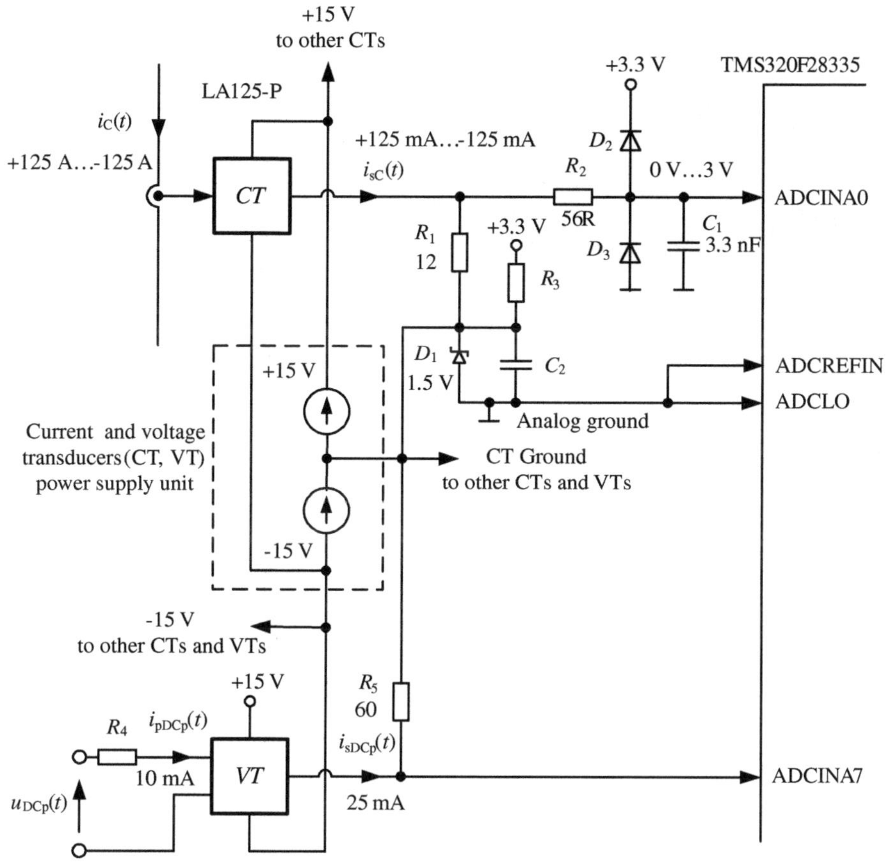

Fig. 4.52 Simplified diagram of galvanically isolated analog input circuit

The APF circuit contains DC energy storage, provided by two capacitors C_1 and C_2. The control circuit is realized using the floating-point digital signal processor TMS320F28335. The control APF algorithm is depicted in Fig. 4.51.

4.4.5.2 Analog Input Circuit

The APF has eleven analog input signals: three from load currents i_{L1}, i_{L2}, i_{L3}, six from inverter output currents i_{Cs1}, i_{Cs2}, i_{Cs3}, i_{Cf1}, i_{Cf2}, i_{Cf3} and two from DC capacitor voltages u_{DCp}, u_{DCn}. All signals are sampled with a sampling rate equal to $f_s = 102{,}400$ Hz. For the electronic measurement of currents, with a galvanic isolation between the primary circuit (high power) and the secondary circuit (electronic circuit) current transducers LEM LA 125-P are used. Primary current is transformed to secondary current according to conversion ratio K_N. For primary

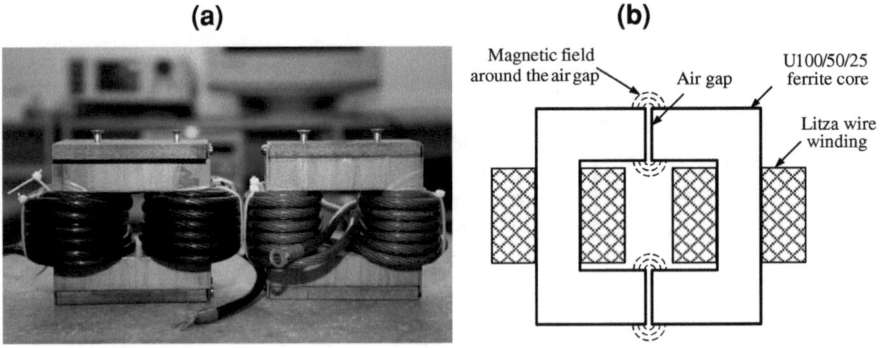

Fig. 4.53 Inductors: **a** the view of inductors, **b** location of windings

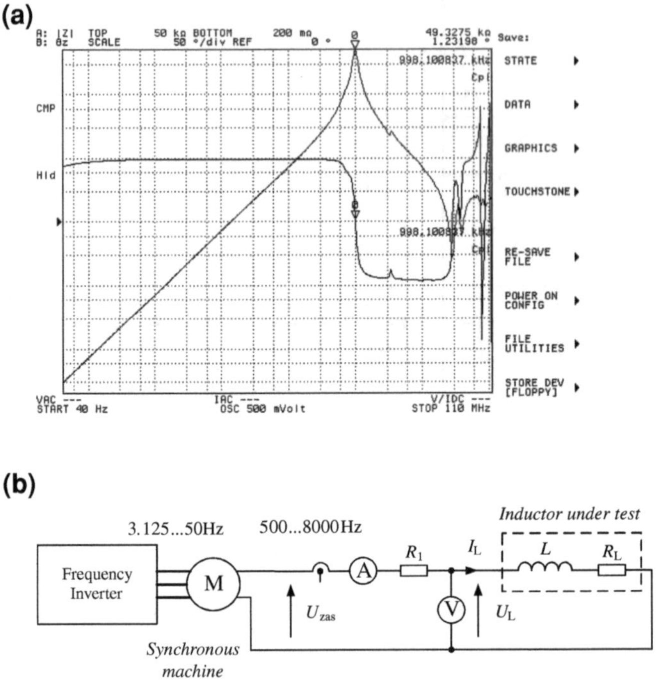

Fig. 4.54 L_{Cf} Inductor test: **a** small signal frequency response, **b** diagram of the inductor test circuit

current $i_{pCT} = 125$ A the secondary is equal to $i_{sCT} = 125$ mA. In a similar way to the currents the voltages are also transformed. Primary voltage is transformed to current signal and next to secondary current. In the considered circuit LEM voltage transducers LV25-P are used. For primary current $i_{pVT} = 10$ mA the secondary is equal to $i_{sVT} = 25$ mA.

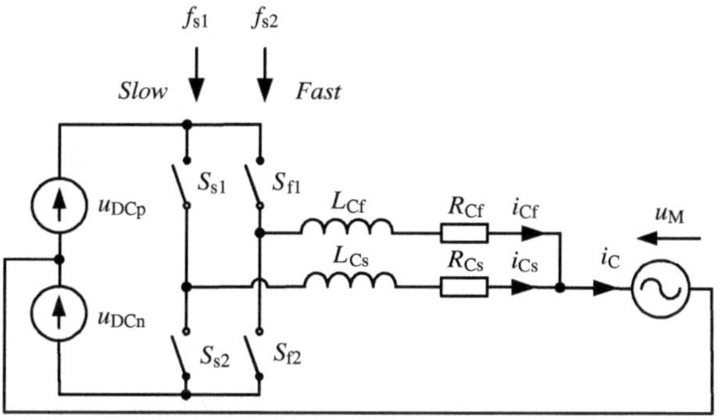

Fig. 4.55 Simplified diagram of modified inverter model connected to the power line

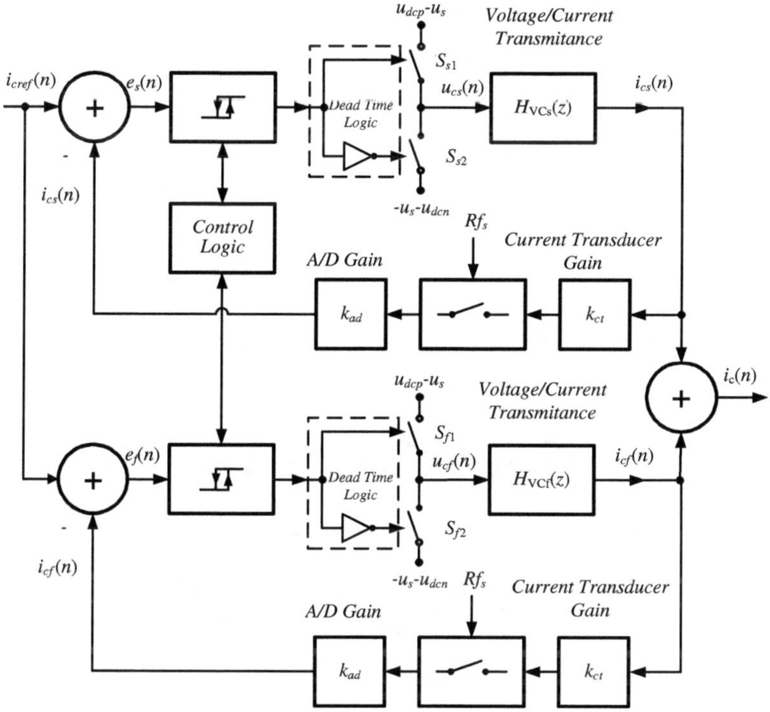

Fig. 4.56 Simplified block diagram of the output inverter simulation circuit

The TMS320F28335 has a 12-bit A/D converter with voltage input range equal to 0–3 V. Unfortunately these inputs are unipolar, therefore all analog input signals have to be transformed to this range. Figure 4.52 shows a simplified diagram of the analog input circuit. ADC input voltage for current transducer can be calculated by the formula:

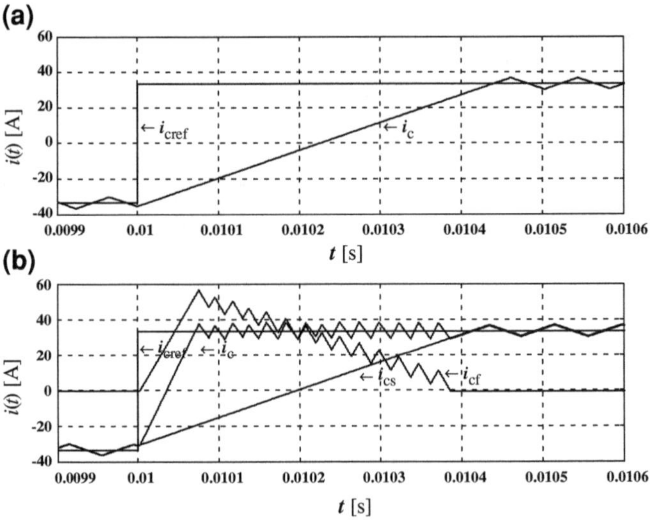

Fig. 4.57 Step responses of two inverters: **a** classical inverter $i_{cref}(t)$, $i_c(t)$, **b** modified inverter $i_{cref}(t)$, $i_c(t)$, $i_{cf}(t)$, $i_{cs}(t)$

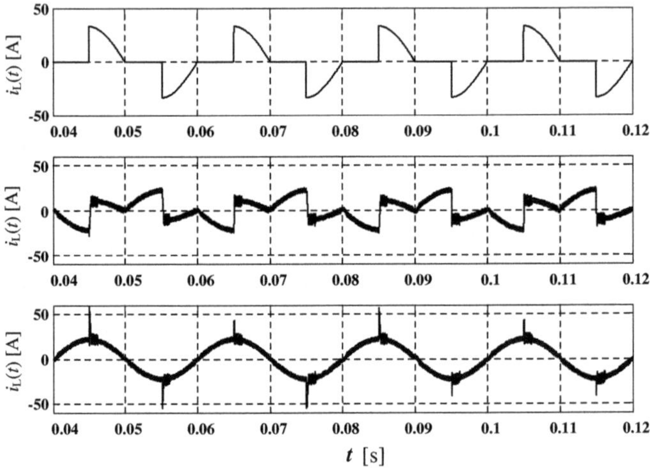

Fig. 4.58 Simulation waveforms of single-phase active power filter in steady-state with the resistive load, with modified output inverter: load current i_L, compensating current i_C, line current i_M

$$U_{ADCIN0} = \frac{I_C}{K_{NCT}} R_1 - U_{D1} \tag{4.54}$$

where: K_{NCT}—current transfer ratio of current transducer, and for voltage transducer:

$$U_{\mathrm{ADCIN7}} = \frac{U_{\mathrm{DCp}}}{R_4} \frac{1}{K_{\mathrm{NVT}}} R_5 - U_{D1} \tag{4.55}$$

4.4.5.3 The Output Inductors

The output inductor has to be linear across the whole output current range. Another important inductor parameter is frequency characteristics of inductor value. With respect to high frequency modulation components generated by the output inverter, the inductor value has to be linear over a frequency range. Coil design considerations are described in Bossche and Valchev [43]. In the considered application the value of inductors was chosen equal [37, 38] to: $L_{\mathrm{Cf}} = 0.5$ mH and $L_{\mathrm{Cs}} = 2.5$ mH. To reduce high-frequency losses, ferrite cores U100/50/25 with air gaps and litz wire were chosen. The inductor windings are located within some distance from the air gap to avoid an induction heating of the windings by the transverse field at that place. The picture of the designed inductors L_{Cf}, L_{Cs} is shown in Fig. 4.53a. Location of the windings is depicted in Fig. 4.53b. Small signal frequency responses of the designed inductors were checked using Agilent 4294A precision impedance analyzer. The result of these measurements is shown in Fig. 4.54a.

The frequency response of inductor impedance is linear up to 1 MHz. Additionally, inductor parameters were checked using high power sinusoidal voltage source (Fig. 4.54b) with regulated output 0–500 V/50 A, and frequency range 500–8,000 Hz. This voltage source was designed by the author using an old synchronous machine formerly used for induction heating.

4.4.5.4 APF Simulation Results

At the beginning a hysteresis digital modulator was designed for controlling the modified inverter. Taken into consideration during the simulation analysis were the modified inverter and classical inverter. The simulation parameters are: $L_{\mathrm{Cf}} = 0.5$ mH, $L_{\mathrm{Cs}} = 2.5$ mH, $u_{\mathrm{DC}} = 390$ V, $f_{s2} = 102{,}400$ Hz, $f_{s1} = 25{,}600$ Hz. Simplified diagram of the modified inverter simulation circuit is shown in Fig. 4.55. The control algorithm of two hysteresis digital modulator with additional conditional control logic implemented in Matlab is shown in Fig. 4.56. Step responses for modified inverter and classic inverter are shown in Fig. 4.57. The classic inverter response time is about 420 μs, and is near 70 μs for the modified inverter. The hysteresis digital modulator is one of the simplest and safest, especially at the early experimental stage, but it has a lot of disadvantages, especially for digital implementation, therefore during future investigations other modulator control algorithms will be designed and implemented.

Figure 4.58 shows the simulation waveforms for a circuit with modified inverter. The following waveforms are depicted: load current i_{L}, compensating current i_C, line current i_M. Using the modified inverter it is possible to decrease the harmonic contents in power line currents from about THD = 15% to about THD = 5%. The results of the simulation analysis confirm good dynamic performance of the modified inverter used in a shunt active power filter.

For predictable nonlinear loads which vary slowly compared to line voltage period (rectifiers, motors etc.) it is easier to predict current changes. For such loads, by using a shunt active power filter with a predictive (non-causal) algorithm, it is possible to decrease harmonic contents. However, for noise type nonlinear loads (such as in an arc furnace), where the load current changes are nonperiodic and stochastic, the proposed APF with improved dynamic performance is a good solution.

References

1. Mohan N, Undeland TM, Robbins WP (1995) Power electronics converters, applications and design, 2nd edn. Wiley, New York
2. Erickson RW, Maksimovic D (2004) Fundamentals of power electronics, 2nd edn. Kluwer Academic Publishers, Dordrecht
3. Bose BK (2006) Power electronics and motor drives: advances and trends. Academic, Burlington
4. Trzynadlowski A (2010) Introduction to modern power electronics. Wiley-Interscience, New York
5. Kazimierkowski M, Malesani L (1998) Current control techniques for three-phase voltage-source converters: a survey. IEEE Trans Ind Electron 45(5):691
6. Kazimierkowski M, Kishnan R, Blaabjerg F (2002) Control in power electronics. Academic, New York
7. Buso S, Mattavelli P (2006) Digital control in power electronics. Morgan and Claypool, Princeton
8. Crochiere RE, Rabiner LR (1983) Multirate digital signal processing. Prentice Hall, Englewood Cliffs
9. Vaidyanathan PP (1993) Multirate systems and filter banks. Prentice-Hall, Upper Saddle River
10. Fliege NJ (1994) Multirate digital signal processing. Wiley, Chichester
11. Nyquist H (1924) Certain factors affecting telegraph speed. Bell Syst Tech J 3:324–346
12. Nyquist H (1928) Certain topics in telegraph transmission theory. AIEE Trans 47:617–644
13. Hartley RVL (1928) Transmission of information. Bell Syst Tech J 7:535–563
14. Shannon CE (1948) A mathematical theory of communication. Bell Syst Tech J vol 27:379–423 pp 623–656 Oct 1948
15. Kotelnikov VA (1933) On the capacity of the 'ether' and of cables in electrical communication. In: Proceedings of the first all-union conference on the technological reconstruction of the communications sector and low-current engineering, Moscow
16. Oppenheim AV, Schafer RW (1989) Discrete-time signal processing. Prentice Hall, New Jersey
17. Proakis JG, Manolakis DM (1996) Digital signal processing, principles algorithms and application. Prentice Hall Inc, Englewood Cliffs
18. Lyons R (2004) Understanding digital signal processing, 2nd edn. Prentice Hall, Englewood Cliffs
19. Rabiner LR, Gold B (1975) Theory and application of digital signal processing. Prentice Hall Inc, Englewood Cliffs
20. Data Translation (2008) The battle for data fidelity: understanding the SFDR spec. Technical report, data translation

21. Data Translation (2009) Benefits of simultaneous data acquisition modules. Technical report, data translation
22. Kester W (2004) Analog–digital conversion, Analog Devices
23. Kester W (2005) The data conversion handbook. Newnes, London
24. Kester W (2009) Understand SINAD, ENOB, SNR, THDTHD, THD + N, and SFDR so you don't get lost in the noise floor. Mt-003, Tutorial, Analog Devices Inc
25. Brannon B, Barlow A (2006) Aperture uncertainty and ADC system performance. Application note, AN-501, Analog Devices Inc
26. Texas Instruments (2008) TMS320F28335/28334/28332, TMS320F28235/28234/28232, Digital signal controllers (DSCs). Data Manual, Texas Instruments
27. Analog Devices (2009) SHARC Processor ADSP-21367, ADSP-21368, ADSP-21369. Data sheet, Analog Devices, Rev E
28. Strzelecki R, Fedyczak Z, Sozański K, Rusiński J (2000) Acitve power filters EFA1, technical report. Instytut Elektrotechniki Przemysłowej, Politechnika Zielonogórska (in polish)
29. Sozański K, Strzelecki R, Kempski A (2002) Digital control circuit for active power filter with modified instantaneous reactive power control algorithm. In Conference proceedings: IEEE 33rd annual IEEE power electronics Specialists conference—PESC '02: conference proceedings, Cairns, Australia, Piscataway, pp 1031–1036
30. Akagi H, Kanazawa Y, Nabae A (1984) Instantaneous reactive power compensators comprising switching devices without energy storage components. IEEE Trans Ind Appl 1A-20(3):625–630
31. Akagi H (1996) New trends in active filters for power conditioning. IEEE Trans Ind Appl 32(6):1312–1322
32. Akagi H, Watanabe EH, Aredes M (2007) Instantaneous power theory and applications to power conditioning. Wiley-Interscience, New Jersey
33. Mariethoz S, Rufer A (2002) Open loop and closed loop spectral frequency active filtering. IEEE Trans Power Electron 17(4):564–573
34. Marks J, Green T (2002) Predictive transient-following control of shunt and series active power filters. IEEE Trans Power Electron 17(4):574–584
35. Wojciechowski D, Strzelecki R (2007) Sensorless predictive control of three-phase parallel active filter. In: Conference proceedings: AFRICON 2007, Windhoek
36. Sozański K (2004) Non-causal current predictor for active power filter. In: Conference proceedings: Nineteenth annual IEEE applied power electronics conference and exhibition, APEC 2004, Anaheim, USA
37. Sozanski K (2006) Sliding DFT control algorithm for three-phase active power filter. 21rd Annual IEEE applied power electronics conference—APEC '06, Dallas, Texas, USA
38. Sozański K (2007) The shunt active power filter with better dynamic performance. In: Conference proceedings: Power tech 2007 conference, Lausanne, Switzerland
39. Sozański K (2008) Shunt active power filter with improved dynamic performance. In: Conference proceedings: 13th international Power electronics and motion control conference—EPE-PEMC 2008, Poznań, Polska: 2018–2022
40. Sozański K (2008) Improved shunt active power filters. Przegląd Elektrotechniczny (Electr Rev) 45(11):290–294
41. Sozański K (2011) Control circuit for active power filter with an instantaneous reactive power control algorithm modification, Przegląd Elektrotechniczny (Electrical Review), nr 1:95–113
42. Jacobsen E, Lyons R (2003) The sliding DFT. Signal Processing Magazine IEEE 20(2): 74–80
43. Van den Bossche A, Valchev VC (2005) Inductors and transformers for power electronics. CRC Press, Boca Raton

Chapter 5
Control and Application of Parallel Active Compensators

Marcin Jarnut and Grzegorz Benysek

Abstract Parallel Active Power Compensators (APC), their topologies and control methods are the major theme of this chapter. The material introduces a different point of view than the usual one on parallel active compensator structure and mode of operation. The range of realized functions in the field of power conditioning is the most important factor taken into consideration. In particular, the comparison of properties in two modes of operations is discussed.

5.1 Introduction

Parallel Active Power Compensators (APC) seem to have been a very widely discussed matter of many publications in the last 20 years [1–7]. The features of these devices can be considered in respect to a few aspects, such as power stage structure, reference current calculation and control method, overall cost of application, number of functions. All of these elements are very important in the development process and are often dependent on each other. Many types of control methods for power compensators have been investigated up to now [5, 8–27]. Significantly, most of them are based on load current decomposition into active

---

M. Jarnut · G. Benysek (✉)
Institute of Electrical Engineering, University of Zielona Góra,
50 Podgórna Street, 65-246 Zielona Góra, Poland
e-mail: G.Benysek@iee.uz.zgora.pl

M. Jarnut
e-mail: m.jarnut@iee.uz.zgora.pl

G. Benysek and M. Pasko (eds.), *Power Theories for Improved Power Quality*,
Power Systems, DOI: 10.1007/978-1-4471-2786-4_5,

Table 5.1 Detection methods for APF control [18]

Calculation domain	Detection method
Time domain [14]	Synchronous fundamental "d-q frame"
	Synchronous individual "d-q frame"
	Instantaneous power
	Generalized integrators and averaging
Frequency domain [22]	Fast Fourier Transform (FFT)
	Discrete Fourier Transform (DFT)
	Recursive Discrete Fourier Transform (RDFT)

and non-active components using instantaneous power theory [10, 12, 15, 17], Fryze's orthogonal currents theory [13, 27] or the newest digital ones such as DFT and RDFT [23–26] (see Table 5.1). The power theory sources of these methods have been described in Chap. 2.

Although the various methods seem to be sometimes quite different, because of calculations in different domains, they have a common idea which is to draw the sine wave line current into phase with the line voltage.

All of the methods in the compensation process require three steps (see Fig. 5.1):

- *step* 1: measurement (sensors);
- *step* 2: detection (processing units);
- *step* 3: regulation (controllers and modulators)

It should be noted that all steps in each method, besides having a different algorithm, employ hardware tools with a different number, type and sensitivity of sensors, different power calculation and calculation speed and, finally, different accuracy of control.

Active compensators controlled with a strategy based on instantaneous power theory and DFT could give quite a fast time response in the case of load current changes, but they need more measurements than compensators based on averaging methods, and the detection circuit requires expensive, more effective, processing tools in these cases. This is especially the case in single phase systems, where the total cost of implemented compensators is the most significant factor.

The number of implemented functions is another important element which plays a role in the process of selecting the most suitable application for given compensation conditions and aims. Recently, according to the rule of "one investment, many benefits", there has been a trend to integrate many functions, such as, uninterruptible power supply (UPS), current compensation and voltage stabilization into one multifunctional power electronics solution [28–34]. The simple APC based on bidirectional pulse width modulated AC/DC converters intended previously only for current compensation became a powerful tool for the regulation of many parameters, such as power flow [1]. Sometimes the realization

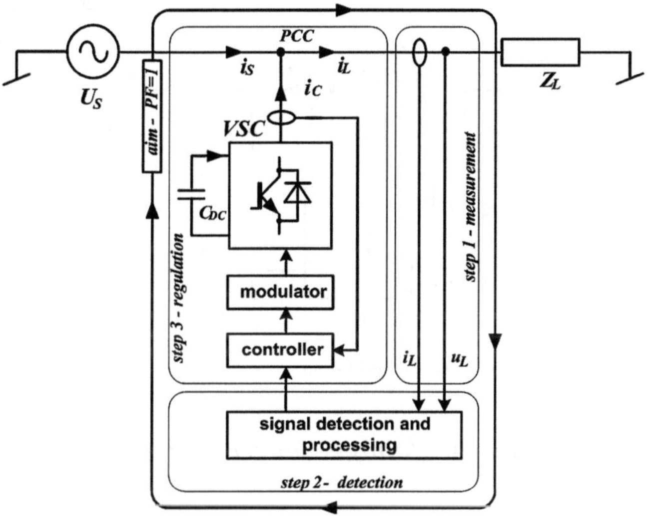

Fig. 5.1 Compensation process with parallel APC

of other functions than current compensation needs only additional control loops and measurements, and sometimes it requires the reconfiguration of the compensator structure.

The next part of this chapter introduces a quite different approach to parallel compensation. Two types of parallel compensators with the same control method but with different mode of operation are described and compared [33]. The hybrid solution composed of both types—voltage and current mode—is introduced to show the wide range of possibilities of modern power electronics applications.

5.2 Active Power Compensators Based on Average Power Method

A typical compensating system (Fig. 5.2) uses a power electronics converter connected to the load terminals at the point of common coupling (PCC). The main functional aim of such a compensating system configuration is to secure a pure sine wave line current i_S. The second assumption about all active compensators is a non-active power flow in the parallel leg, which means $P_C = 0$.

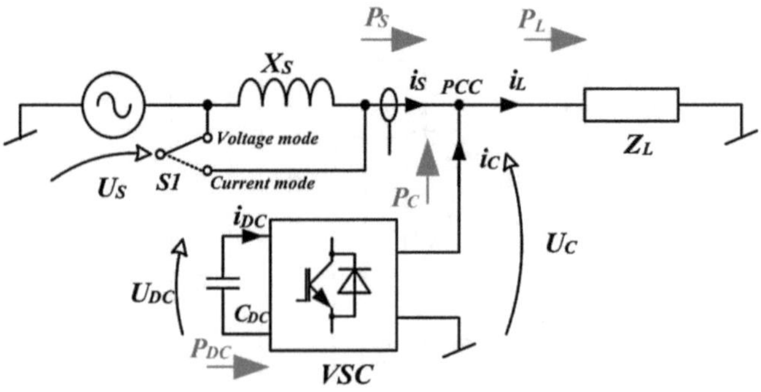

Fig. 5.2 Simplified schema of power conditioner

5.2.1 Power Balance in Voltage Source Compensator

For proper operation compensators based on voltage source converters (VSC) require DC link voltage stabilization. In the case where a conditioner does not have an external energy source connected to the DC terminals, a voltage stabilization loop is needed in its control unit. Energy changes on the DC storage element C_{DC} always cause voltage fluctuations at its terminals:

$$E_{(0)} - E_{(\mathrm{T})} = C_{\mathrm{DC}} \cdot \frac{U^2_{\mathrm{DC}(0)} - U^2_{\mathrm{DC}(\mathrm{T})}}{2} \tag{5.1}$$

Power and energy relations in time interval T_C are described thus:

$$E_{(0)} - E_{(\mathrm{T})} = P_{\mathrm{DC}} \cdot T_C \tag{5.2}$$

and for relatively small U_{DC} voltage variations this means:

$$P_{\mathrm{DC}} = U_{\mathrm{DC}(0)} \cdot C_{\mathrm{DC}} \cdot \frac{U_{\mathrm{DC}(0)} - U_{\mathrm{DC}(T)}}{T_C} \tag{5.3}$$

Based on above it can be concluded that compensator average power P_{DC} is a function of the DC link voltage variations in averaging period T_C. Based on the assumption that converter lossless $P_{\mathrm{loss}} = 0$, the compensator AC and DC side average powers are equal $P_{\mathrm{DC}} = P_C$. To satisfy the voltage stabilization rules described above the voltage source converter should control the power balance at the PCC:

$$P_S + P_C = P_L \xrightarrow{P_C=0} P_S = P_L \tag{5.4}$$

The P_L load power in the case of a constant load voltage in PCC is independent of compensator activity, hence the inverter can control only source power P_S. In this case from the DC link voltage U_{DC} stabilization point of view, the loads are visible as disturbances, as shown in Fig. 5.3.

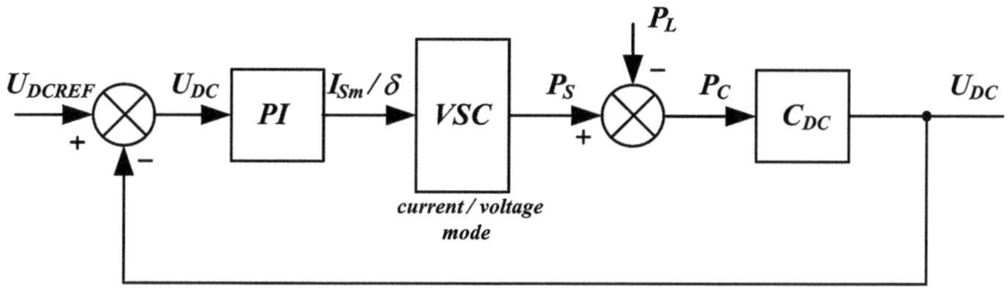

Fig. 5.3 Simplified schema of power balance method

Usually the control circuit in averaging methods needs at least one averaging period T_C to respond to disturbance changes. In this time interval, the inverter supplies the loads and this causes DC link capacitor C_{DC} voltage fluctuations ΔU_{DC} which can be expressed as the function of the load's active power changes ΔP_L:

$$\Delta U_{\mathrm{DC}} = \frac{\Delta P_L \cdot T_C}{U_{\mathrm{DCREF}} \cdot C_{\mathrm{DC}}}, \tag{5.5}$$

where U_{DCREF} is the reference value of DC link voltage.

It should be noted that the above relations give reliable results only if voltage fluctuations caused by load power changes are much smaller than the DC link reference voltage. This can be obtained by sizing the DC link capacitance to give small voltage drops between capacitor terminals (usually less than 5%).

DC link voltage drop generates growth of the voltage stabilization error and further growth of the reference current magnitude I_{Sm} or the converter's reference voltage phase angle signal δ.

Line active power P_S increases which is proportional to the load's active power growth (Fig. 5.2), and can be achieved using two types of power balance modes: current mode and voltage mode [30–32] (Fig. 5.4).

5.2.2 Current Mode Power Balance

Current type compensators are the most popular solution for power compensators. In this arrangement the inverter produces current i_C which is equal to the non-active component in the load current i_{Ln}. It compensates the current at the PCC. Then the source current i_S contains only an active component i_{Sa}.

According to Kirchhoff's law:

$$i_S = i_L - i_C \tag{5.6}$$

and Fryze theory [27]:

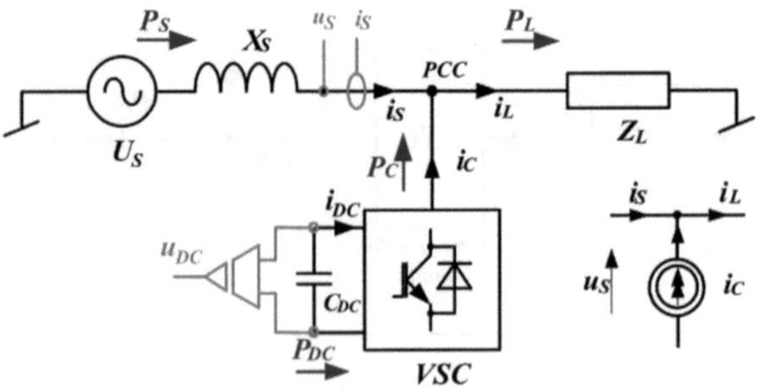

Fig. 5.4 Current mode APC schema

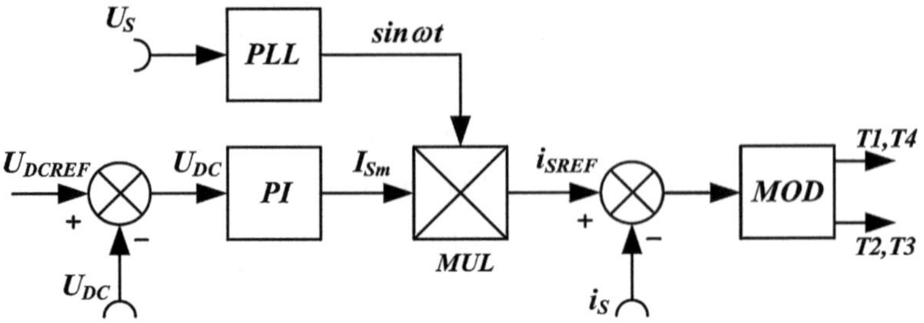

Fig. 5.5 Control schema of current-type conditioner

$$
\begin{aligned}
i_S &= i_{\mathrm{Sa}} + i_{\mathrm{Sn}} \\
i_L &= i_{\mathrm{La}} + i_{\mathrm{Ln}} \\
i_C &= i_{\mathrm{Ca}} + i_{\mathrm{Cn}}
\end{aligned} \tag{5.7}
$$

and taking into consideration DC link voltage stabilization $P_C = 0$:

$$i_{\mathrm{Ca}} = 0 \tag{5.8}$$

the compensation equations can be expressed:

$$i_S = i_{\mathrm{Sa}} = i_{\mathrm{La}} \Leftrightarrow i_{\mathrm{Cn}} = i_{\mathrm{Ln}} \tag{5.9}$$

Using this short analysis, the source of the reference current can be defined as:

$$i_{\mathrm{SREF}} = I_{\mathrm{Sm}} \cdot \sin \omega t, \tag{5.10}$$

where I_{Sm} is the magnitude of source active current i_{Sa}.

From the above equations one can conclude that the DC link voltage stabilization, together with the shaping of the source current to contain an active component, is enough to compensate a load's non-active current component. A control unit diagram based on these rules is shown below (Fig. 5.5).

The magnitude of the source active current is calculated in the DC link voltage stabilization loop and the reference waveform is drawn in the PLL unit which is synchronized with the source voltage u_S. Because at the PCC we do have the same voltage for all components—source, compensator and load—current balance equals the active power balance as well. The source active power is then:

$$P_S = U_S \cdot I_S, \tag{5.11}$$

where U_S and I_S are the RMS values of the source voltage u_S and current i_S.

Using (5.4), (5.5) and (5.11) the proportional gain in *PI* controller can be determined as follows:

$$k_{\mathrm{I}} = \frac{I_{\mathrm{Sm}}}{\Delta U_{\mathrm{DC}}} = \frac{\sqrt{2} \cdot C_{\mathrm{DC}} \cdot U_{\mathrm{DCREF}}}{U_S \cdot T_C} \tag{5.12}$$

The integrating time T_{I} of this controller should not be lower than T_C time, which is an averaging interval of the P_C power. Otherwise the control algorithm could give faulty reference signals.

5.2.3 Voltage Mode Power Balance

In the voltage mode of operation the compensator shapes the u_C voltage at the PCC. This voltage has the same parameters as the nominal source voltage u_S but is shifted by δ angle in relation to this voltage:

$$u_C = U_{\mathrm{Cm}} \cdot \sin(\omega t - \delta), \tag{5.13}$$

where U_{Cm} is nominal load voltage magnitude (Fig. 5.6).

The compensation of line current then proceeds "naturally" on the assumption that the difference of two sinusoidal voltages (U_S, U_C) gives the sinusoidal current. The phase angle δ is calculated in the DC link voltage stabilization loop and should be a function of load active power P_L, as shown in Fig. 5.2.

Synchronization is determined in the PLL block, which generates signal proportional to source voltage phase angle ωt (Fig. 5.7).

To simplify the control strategy the assumption is made that the range of possible δ changes is 0–45°. This assumption allows the linearization of dependency $\delta/\sin\delta$, which can be expressed in this range as the proportional gain, having a value of about 57. In further considerations, taking into account this assumption, the *PI* controller gain is as follows:

$$k_U = \frac{\delta}{\Delta U_{\mathrm{DC}}} = \frac{57 \cdot C_{\mathrm{DC}} \cdot X_S \cdot U_{\mathrm{DCREF}}}{U_S \cdot U_C \cdot T_C} \tag{5.14}$$

The voltages at both inductance terminals have sine wave shapes, hence the current in this element will also be sinusoidal. Possible line voltage distortion usually does not exceed 5% and does not increase distortions in any significant

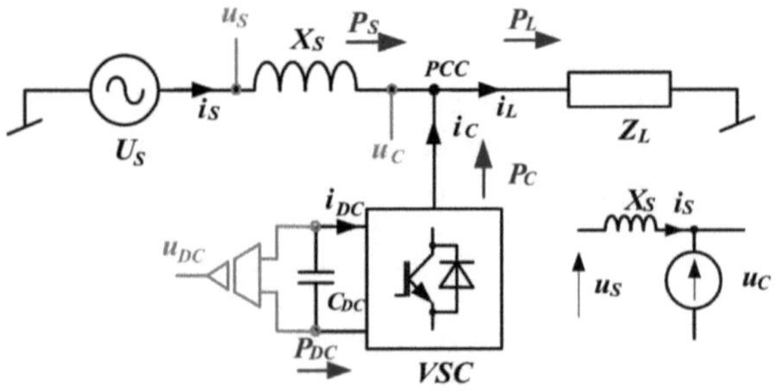

Fig. 5.6 Voltage mode APC schema

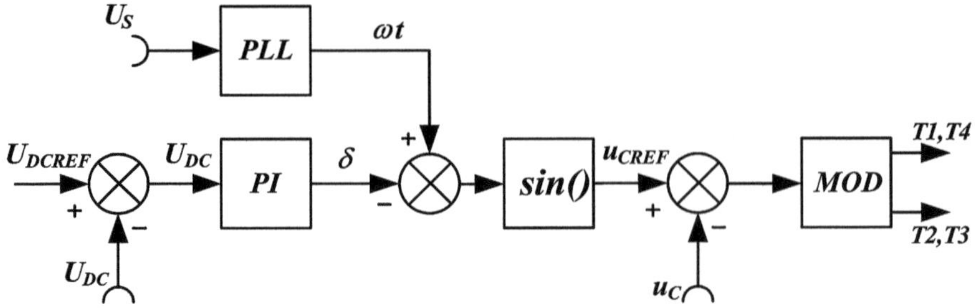

Fig. 5.7 Control schema of voltage-type compensator

way in the inductance current. Voltage shaping at the output of inverter needs a DC link voltage stabilization, which is possible only if the compensator does not exchange active power with the system. This means that $P_C = 0$, which is possible only if $P_S = P_L$. The active power in X_S element is controlled by load voltage u_C and phase angle δ changes:

$$P_S = \frac{U_S \cdot U_C}{X_S} \cdot \sin\delta \tag{5.15}$$

It should be noted that in satisfying the DC link voltage stabilization equations and load voltage shaping the compensator can be controlled without any current measurements.

5.2.4 Mixed Mode Power Balance

Voltage mode parallel active compensators have one significant disadvantage: the power factor depends on the load's active power and line voltage. This causes *PF* deterioration, especially in the case of line voltage dips and swells (although the load voltage in PCC still is stable). Integration of the advantages of both voltage and current modes into one parallel–parallel structure gives the opportunity to build a very effective power compensating system [35] (Fig. 5.8).

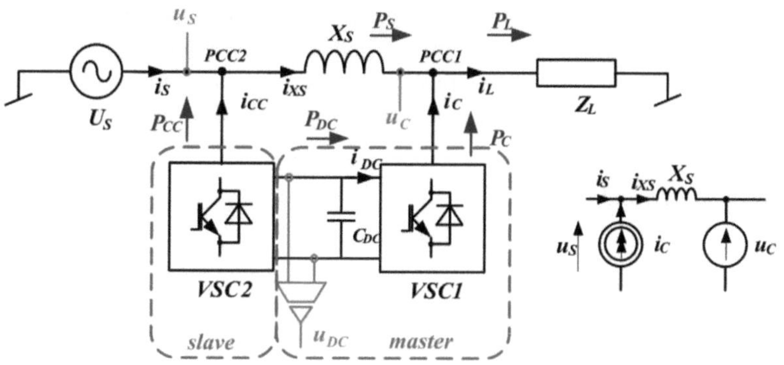

Fig. 5.8 Mixed mode APC schema

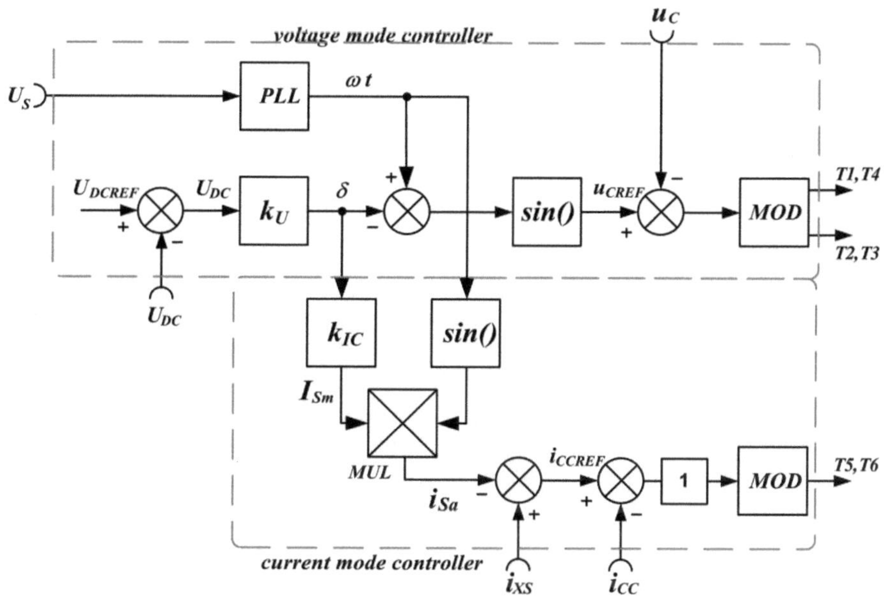

Fig. 5.9 Control unit of mixed mode parallel–parallel compensator

The structure of the compensating system can be divided into two parts: in the first one the slave is working in current mode; in the second one, the master is working in voltage mode. This division allows the elimination of the disadvantages of the voltage part, while still maintaining all of its advantages. Moreover, in single phase solutions the "ordinary" three leg inverter can be implemented as a mixed mode parallel–parallel compensator. It presents the opportunity for cost reduction, with the same functionality and effectiveness (Fig. 5.9).

Although the voltage mode part works as described above, the slave part realizes only line current compensation (power factor compensation) using signals from the control unit of the master part and one additional signal from the series inductor current to satisfy the current compensation equations at point PCC2:

- according to Kirchhoff law:

$$i_S = i_{\mathrm{XS}} + i_{\mathrm{CC}} \tag{5.16}$$

- using Fryze theory:

$$\begin{aligned} i_S &= i_{\mathrm{Sa}} + i_{\mathrm{Sn}} \\ i_{\mathrm{XS}} &= i_{\mathrm{XSa}} + i_{\mathrm{XSn}} \\ i_{\mathrm{CC}} &= i_{\mathrm{CCa}} + i_{\mathrm{CCn}} \end{aligned} \tag{5.17}$$

taking into consideration DC link voltage stabilization $P_{\mathrm{CC}} = 0$:

$$i_{\mathrm{CCa}} = 0 \tag{5.18}$$

compensation equations can be expressed as follows:

$$i_S = i_{\mathrm{Sa}} = i_{\mathrm{XSa}} \Leftrightarrow i_{\mathrm{CCn}} = i_{\mathrm{XSn}} \Rightarrow i_{\mathrm{CCn}} = i_{\mathrm{XS}} - i_{\mathrm{Sa}} \tag{5.19}$$

The active component in line current can be determined using δ angle which is the function of P_L load average power, according to (5.4) and (5.14).

It should be noted that from the compensation point of view the slave part is controlled in open loop. It is very appropriate in such a structure, because there is only one closed loop responsible for DC voltage stabilization. Thus the compensator remains stable despite changing line voltage and load power parameters. Because the current mode part depends on the voltage mode, then the k_{IC} controller gain using (5.12) and (5.14) can be described as follows:

$$k_{\mathrm{IC}} = \frac{I_{\mathrm{Sm}}}{\delta} = \frac{\sqrt{2} \cdot U_C}{57 \cdot X_S} \tag{5.20}$$

The second control loop in the current mode controller allows the drawing of the requested output current using the voltage source converter. The gain of this loop is set to unity.

The major aim of designing parallel–parallel compensation systems is the implementation into one structure as many functions as possible. This is well-established in modern distribution systems where the nonlinear loads and external weather phenomena deteriorate power quality (PQ). Although this is not so in every case of power conditioning, many functions are required, such as power factor correction (PFC), voltage stabilization, UPS or interconnection of distributed generation (DG) with the electrical power system (EPS) (Table 5.2).

Table 5.2 Functions of parallel compensators

Mode of operation/function	Current mode	Voltage mode	Mixed mode
PFC	√	√[a]	√
Voltage stabilization	–	√	√
UPS	–	√[b]	√[b]
DG interconnection	√	√	√

[a] Input power factor near unity
[b] With external energy storage connected to DC link or external energy sources

Table 5.3 Investigated system parameters

	Current mode	Voltage mode	Mixed mode
Source voltage (U_S)	230 V/50 Hz	230 V/50 Hz	230 V/50 Hz
Load voltage (U_C)	230 V/50 Hz	230 V/50 Hz	230 V/50 Hz
Load type	Nonlinear	Nonlinear	Nonlinear
Load power (P_L)	1.5/3 kW	1.5/3 kW	1.5/3 kW
DC link voltage (U_{DCREF})	400 V	400 V	400 V
DC link capacitance (C_{DC})	2 mF	2 mF	2 mF
Series reactance (X_S)	5 mH	5 mH	5 mH
Switching frequency (f_S)	10 kHz	10 kHz	10 kHz
Controller gain (k)	0.27	0.072	3.6/0.072[a]
Controller time constant (T_C)	0.04 s	0.04 s	0.04 s

[a] Current mode controller gain/voltage mode controller gain

5.3 Investigation Results

Three types of compensators with the described control strategies based on the power balance method have been researched using simulation software.

The simulation parameters have been shown in Table 5.3. All types of compensators have been connected to systems with the most common nonlinear load, i.e., the bridge rectifier with capacitive load. The load current distortion $THD(i_L)$ in the considered case is about 85%. The load on the DC side of rectifier has been increased from 1.5 to 3 kW for dynamic property verification. Additionally in a compensating system with mixed mode, a distorted source voltage with voltage sag simulation from 10 to 20% was implemented.

5.3.1 Steady-State Results

The simulation steady-state results shown in Figs. 5.10 and 5.11 prove the compensation effectiveness in both the considered solutions. The distortion factor of the source current in both cases is below the 3% level. In voltage-type arrangements the source current has a little higher magnitude, caused by the reactive

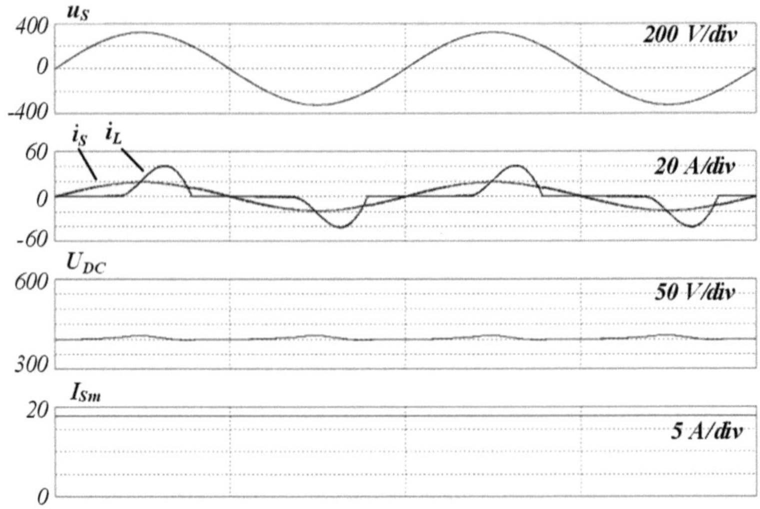

Fig. 5.10 Research results of current mode conditioner

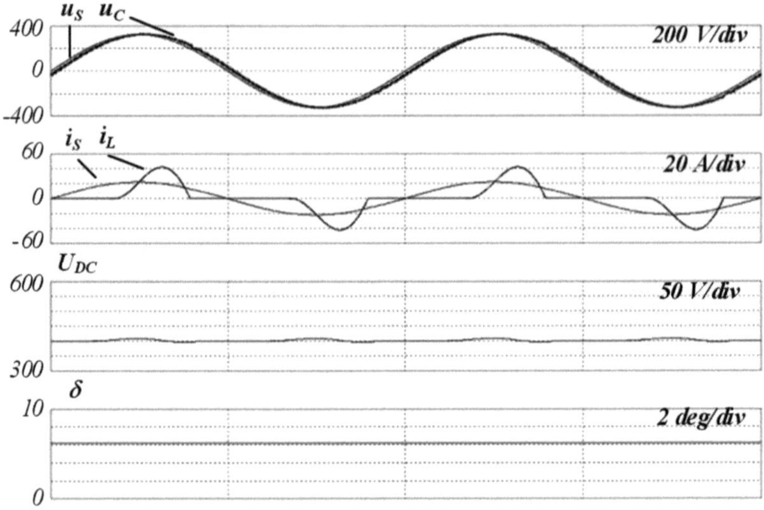

Fig. 5.11 Research results of voltage mode conditioner

power. This reactive power has a small value and the input power factor stays above 0.95. Source current magnitude I_{Sm} seems to be well calculated in current-type compensators and facilitates DC link voltage stabilization and mitigation of load current distortion. The same situation can be observed in voltage-type compensators where an accurately determined value of the load voltage phase angle δ gives the same results.

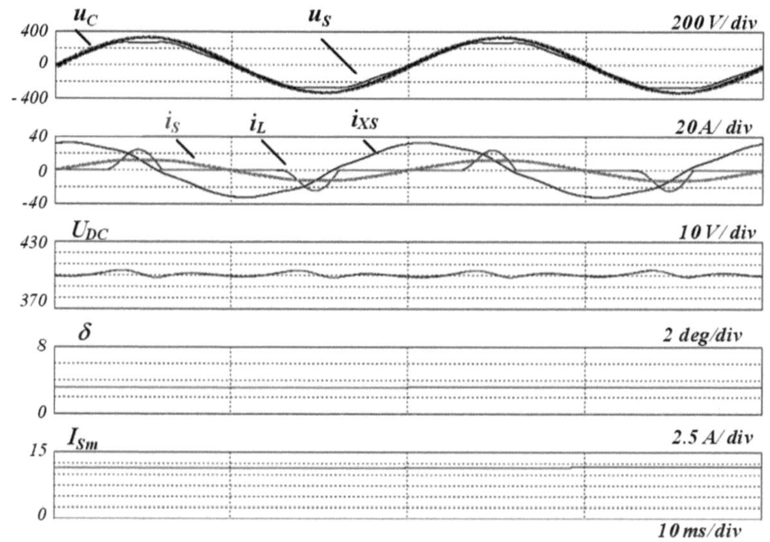

Fig. 5.12 Research results of mixed mode conditioner

In Fig. 5.12 results of a mixed mode compensation system with parallel–parallel structure are shown. As one can see in the second curve in the case when mains voltage is distorted or its magnitude is different to the nominal, the power factor is below 0.5 value (*PF* measured on X_S side of PCC point, which is appropriate to the voltage mode solution). The current mode part of the compensator corrects to unity *PF* (measured between input terminals). Simultaneously the voltage mode part provides the sine wave ($THD(U_C) < 1\%$) voltage with nominal magnitude.

5.3.2 Transients Results

To verify the effectiveness of control the compensators were tested during The solutions were tested during load changes, because the main rule of the control circuit in both cases is the adjustment of source active power P_S to follow up the load active power P_L. The results are presented below.

From Figs. 5.13, 5.14 and 5.15 it can be concluded that in all the considered solutions the controller settings fulfill the chosen regulation criteria of 10% DC link voltage overshoot. After a disturbance the DC link voltage returns to nominal value. Averaging methods give a delayed response as is visible in the depicted results. As will be noticed the parallel–parallel solution gives an even faster response, although its control unit contains only one control loop of the DC link voltage. For improvement of compensation in the simulated

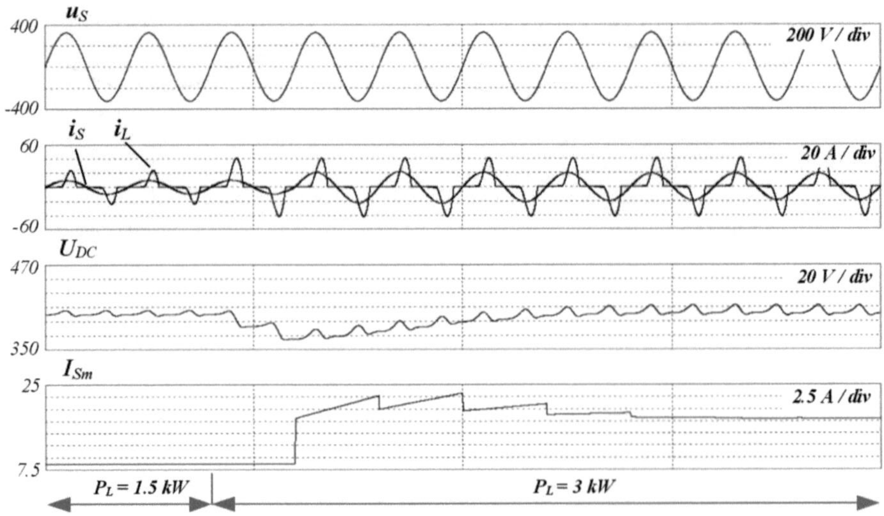

Fig. 5.13 Results of current mode conditioner transients

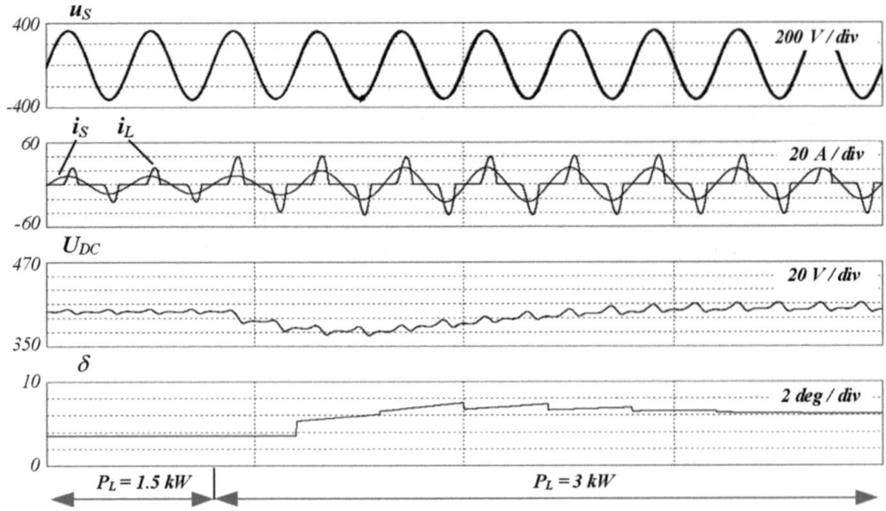

Fig. 5.14 Results of voltage mode conditioner transients

solution zero-order hold (ZOH) has been applied in the DC link voltage sensor circuit. The sampling time of ZOH is 20 ms. Thanks to this the DC link voltage pulsation, caused by non-active power flow, does not influence the

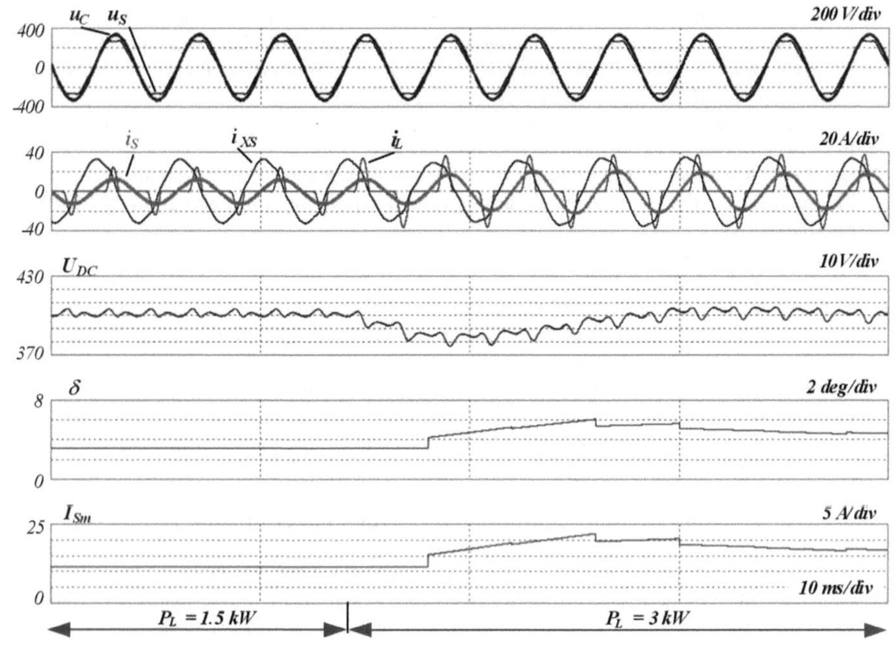

Fig. 5.15 Results of mixed mode conditioner transients

reference values of the source current magnitude and the phase angle of the load voltage. It does not deteriorate the compensator stability in any of the tested cases.

References

1. Hingorani N, Gyugyi L (2000) Understanding FACTS: concepts and technology of flexible ac transmission systems. IEEE, New York
2. Benysek G (2007) Improvement in the quality of delivery of electrical energy using power electronics systems. Springer, London
3. Malesani L, Rossetto L, Tenti P (1986) Active filters for reactive power and harmonic compensation. In: Proceedings of IEEE-PESC, pp 321–330
4. Grady WM, Samotyj MJ, Noyola AH (1990) Survey of active power line conditioning methodologies. IEEE Trans Power Deliv 5(3):1536–1542
5. Akagi H (1996) New trends in active filters for power conditioning. IEEE Trans Ind Appl 32(6):1312–1322
6. Akagi H (1994) Trends in active power line conditioners. IEEE Trans Power Electron 9(3):263–268
7. Strzelecki R (2002) Active arrangements for energy conditioning—APC. (in Polish), Przegląd Elektrotechniczny, no 2:196–202
8. Akagi H, Fujita H (1995) A new power line conditioner for harmonic compensation in power systems. IEEE Trans Power Deliv 10(3):1570–1575

9. Peng FZ, Tolbert LM (2000) Compensation of nonactive current in power systems—definitions from a compensation standpoint. In: Proceedings of PES' 2000, pp 983–987
10. Peng FZ, Lai JS (1996) Generalized instantaneous reactive power theory for three-phase power systems. IEEE Trans Instrum Meas 45(1):293–297
11. Fukuda S, Endoh T (1993) Control method and characteristics of active power filter. In: Proceedings of EPE, Brighton, UK, pp 139–144
12. Soares V, Verdelho P, Marques P (1997) Active power filter control circuit based on the instantaneous active and reactive current id-iq method. In: Proceedings of IEEE-PESC, pp 1096–1108
13. Czarnecki L (1990) A time domain approach to reactive current minimization in nonsinusoidal situations. IEEE Trans Instrum Meas 39:698
14. Tolbert LM, Habetler TG (2000) Comparison of time-based nonactive power definitions for active filtering. In: Proceedings on power electronics congress '00, vol 1, pp 73–79
15. Kim H, Akagi H (1997) The instantaneous power theory based on mapping matrices in three-phase four-wire systems. In: Proceedings of power conversion conference '97, vol 1, pp 361–366
16. Peng FZ, Tolbert LM, Qian Z (2002) Definitions and compensation of nonactive current in power systems. In: Proceedings of PESC'2002, vol 4, pp 1779–1784
17. Kim H, Blaabjerg F, Bak-Jensen B, Jaeho C (2002) Instantaneous power compensation in three-phase systems by using p-q-r theory. IEEE Trans Power Electron 17:701–710
18. Blaabjerg F, Hansen S, Asiminoaei L (2007) Detection is key—harmonic detection methods for active power filter applications. IEEE Ind Appl Mag 13(4):22–33
19. Newman MJ, Zmood DN, Holmes DG (2002) Stationary frame harmonic reference generation for active filter systems. IEEE Trans Ind Appl 38(6):1591–1599
20. Jacobs J, Detjen DO, De Doncker RW (2002) An overview of methods to determine the harmonics in three-phase systems. In: Proceedings of IEEE young researchers symposium electrical power engineering distributed generation
21. Chang GW, Tai-Chang S (2002) A comparative study of active power filter reference compensation approaches. In: Proceedings of PES'02, vol 2, pp 1017–1021
22. Rechka S, Ngandui T, Jianhong X, Sicard P (2002) A comparative study of harmonic detection algorithms for active filters and hybrid active filters. In: Proceedings of PESC'02, vol 1, pp 357–363
23. Williams SM, Hoft RG (1991) Adaptive frequency domain control of PWM switched power line conditioner. IEEE Trans Power Electron 6(4):665–670
24. Solomon OM (1994) The use of DFT windows in signal-to-noise ratio and harmonic distortion computations. IEEE Trans Instrum Meas 43(2):194–199
25. Sozanski K (2004) Harmonic compensation using the sliding DFT algorithm. In: 35th Annual IEEE power electronics specialists conference—PESC '04, Aachen, Germany
26. Borisov K, Ginn H (2009) A computationally efficient RDFT-based reference signal generator for active compensators. IEEE Trans Power Deliv 24:2396–2404
27. Fryze S (1931) Active, reactive and apparent power in non-sinusoidal systems (in Polish). Przeglad Elektrotechniczny 7:193–203
28. Emadi A, Nasiri A, Bekiarov SB (2005) Uninterruptible power supplies and active filters. CRC Press, Boca Raton
29. Chandra A, Singh B, Singh BN, Al-Haddad K (2000) An improved control algorithm of shunt active filter for voltage regulation, harmonic elimination, power factor correction and balancing of nonlinear loads. IEEE Trans Power Electron 15(3):495–507
30. Strzelecki R, Rusinski J, Benysek G (2002) Voltage source power quality conditioner. In: Electromagnetic phenomena in nonlinear circuits–EPNC 2002, XVII Symposium. Leuven, Belgium, pp 179–182
31. Meckien G, Strzelecki R (2002) Single phase active power line conditioners-without transformers. In: EPE–PEMC conference
32. Strzelecki R, Jarnut M, Benysek G (2003) Active electrical energy conditioners for individual customers. In: PES Conference, Warsaw University of Technology Press, vol 1, pp 27–34

33. Strzelecki R, Benysek G, Jarnut M (2008) Power quality conditioners with minimum number of current sensors requirement. Przeglad Elektrotechniczny 11:295–298
34. Strzelecki R, Benysek G, Jarnut M (2007) Interconnection of the customer-side resources using single phase VAPF. Przeglad Elektrotechniczny 10:59–65
35. Strzelecki R, Benysek G, Jarnut M (2005) Symmetrical power line conditioner—basic properties. In: 4th International workshop compatibility in power electronics–CPE 2005, Gdynia, Poland

Chapter 6
Practical Application of Series Active Compensators

Jacek Kaniewski

Abstract This chapter describes series-connected alternating voltage compensators in power systems. The focus of the description is on the structure, operation and control method of serial AC voltage compensators. The main part of this chapter concerns the practical application of the hybrid transformer (HT) (with electrical and electromagnetic coupling) as a special kind of serial AC serial voltage compensator. Presented are the calculations and experimental test results and future directions of the development of these devices.

6.1 Introduction

The parameters of electric energy, and its quality, are very important especially from the viewpoint of the end-user. The parameters of electric energy are well known and described in [1]. According to [1] we can define several voltage parameters such as nominal voltage of the system U_N, declared supply voltage U_D. Classification of the voltage fluctuation depending on size of voltage change is shown in Fig. 6.1.

An expanded and centralized AC electrical system is exposed to many different faults (Fig. 6.2). In the case of dynamic states in an AC electrical system, for example, fast load changes, switching effects, atmospheric discharges, faults, on the consumer side, undesirable effects are generated, such as voltage dips (87.2% of all distortion in the power grid), interruptions and swells [2, 3] (Fig. 6.2).

J. Kaniewski (✉)
Institute of Electrical Engineering, University of Zielona Góra,
50 Podgórna Street, 65-246, Zielona Góra, Poland
e-mail: j.kaniewski@iee.uz.zgora.pl

G. Benysek and M. Pasko (eds.), *Power Theories for Improved Power Quality*,
Power Systems, DOI: 10.1007/978-1-4471-2786-4_6,

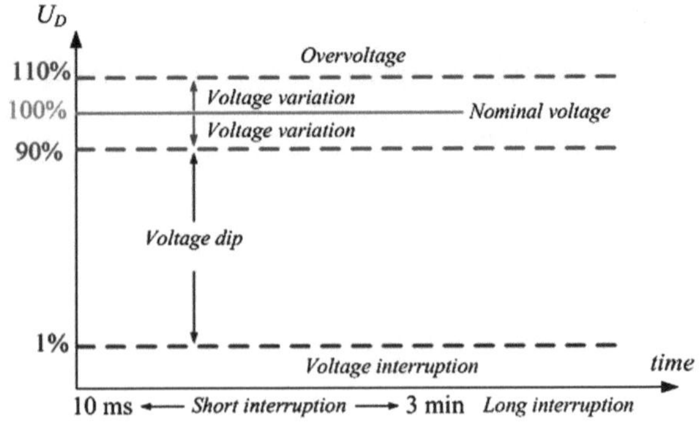

Fig. 6.1 Classification of voltage variation depending on size of change

Poor power quality has a detrimental effect on electric devices. In the case of sensitive devices, erratic supply parameters cause failure or defective devices [4, 5]. In the case of big plants and factories, voltage sags and swells may cause very large financial damage, especially in automotive, pharmaceutical and semiconductors industry [6]. To mitigate the adverse effects of supply voltage the application of voltage sag compensators and voltage regulators is employed, particularly Series Voltage Compensators (SVC). The Operating principle of the SVC is shown in Fig. 6.3 and described in Eq. 6.1.

$$\underline{U}_L = \underline{U}_S \pm \underline{U}_C \tag{6.1}$$

The load voltage U_L is the sum of source voltage U_S and the compensating voltage U_C. During nominal conditions the voltage U_C equals 0 ($U_C = 0$). During source voltage sag the compensating voltage U_C is added to source voltage U_S, and opposite during overvoltage—voltage U_C is subtracted from voltage U_S. A more detailed description of SVC was is found below in this chapter.

Series compensators are well known and are described in [10–19]. An often encountered circuit for the SVC is the dynamic voltage restore (DVR) [12, 15]. The simple schematic diagram of the DVR is shown in Fig. 6.4.

Voltage compensation is based on an injection of a series voltage U_C into the incoming supply voltage U_S during the voltage sag or swell. As can be seen in Fig. 6.4b, c and d, the load voltage U_L is the sum of source voltage U_S and compensating voltage U_C. During nominal conditions the compensating voltage is equal to zero (Fig. 6.4b) (6.2). In the case of a voltage sag in the source voltage U_S the compensating voltage U_C is added in phase with the source voltage U_S (Fig. 6.4c) (6.3). In the case of an overvoltage in the source voltage U_S the compensating voltage U_C is added diametrically out of phase with the source voltage U_S (Fig. 6.4d) (6.4).

$$\underline{U}_S = \underline{U}_N = \underline{U}_L \quad \text{then} \quad \underline{U}_C = 0 \tag{6.2}$$

$$U_S < U_N \quad \text{then} \quad \underline{U}_L = \underline{U}_S + \underline{U}_C \tag{6.3}$$

$$U_S > U_N \quad \text{then} \quad \underline{U}_L = \underline{U}_S - \underline{U}_C \tag{6.4}$$

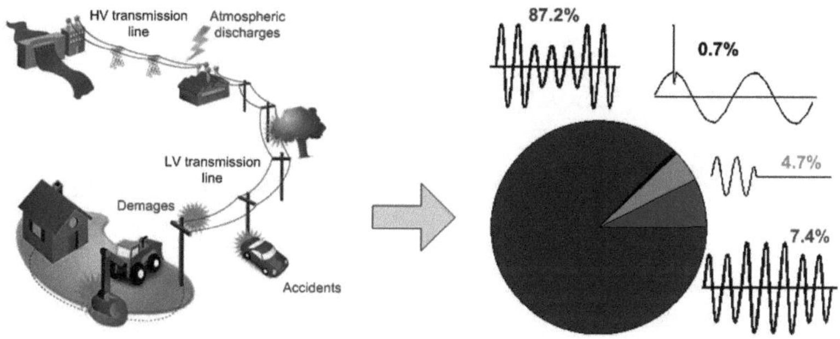

Fig. 6.2 Distortion in power grid—causes and effects

Fig. 6.3 Operating principle of the SVC

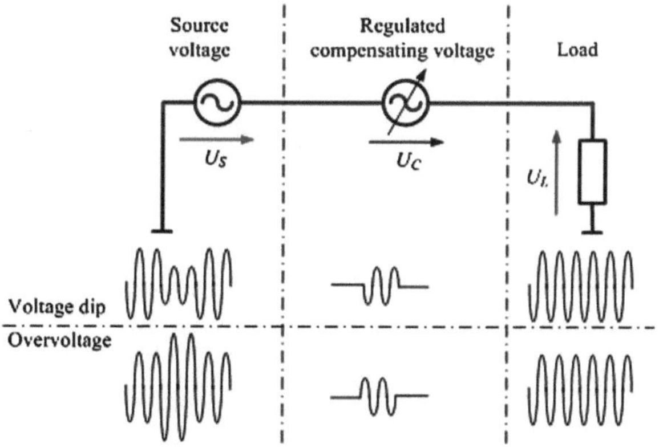

As a DC energy storage unit there are many possibilities, such as batteries, super capacitors, fuel cells and others such as Super Magnetic Energy Storage (SMES).

The energy storage for the supply DC/AC converter makes it possible to compensate deep voltage sags, and even short source voltage interrupts. The operating duration of DVR is dependent on voltage duration depth, capacity of DC energy storage and power of protected load circuit. In the case of deep voltage sag or voltage interrupt, the operating duration of DVR is shorter than in the case of voltage duration around nominal voltage conditions. The limited capacity of DC energy storage is the main disadvantage of DVR. In order to eliminate the disadvantage of the limited capacity of DC energy storage, the SVC with AC/DC/AC converter is used (Fig. 6.5).

As shown in Fig. 6.5 an AC/DC/AC voltage converter is used to supply the series transformer. The voltage converter is supplied directly from the grid (Fig. 6.5a) or by transformer (Fig. 6.5b). This solution makes it possible to compensate long voltage sags and swells.

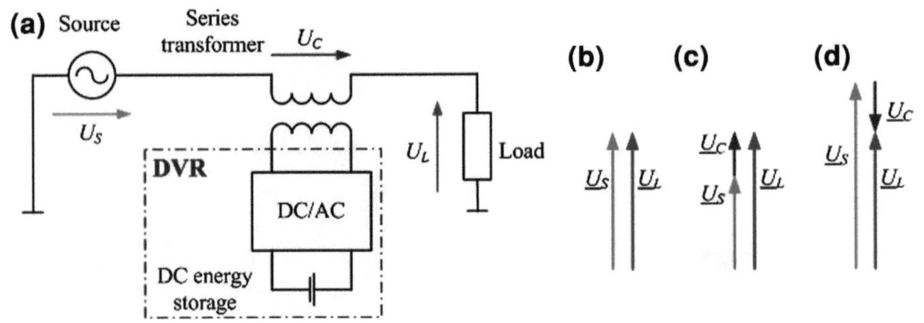

Fig. 6.4 Operating principle of DVR: **a** simplified schematic diagram; voltage phasors in the case of **b** nominal conditions, **c** source voltage dip, **d** overvoltage

Fig. 6.5 Series voltage compensator, **a** supplied directly from grid, **b** supplied by input transformer

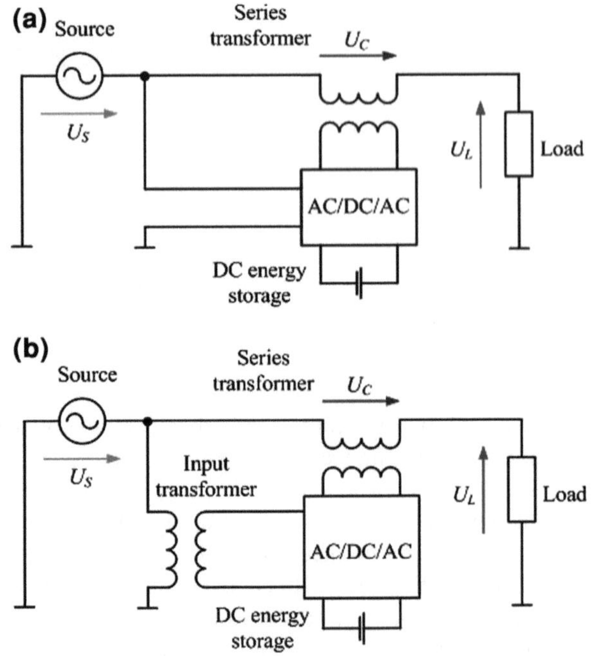

Another special kind of SVC is the topology described in [19] and developed in [8, 9, 20–28]. This solution is based on a combination of an Electromagnetic Transformer (TR) and a PWM AC/AC converter, such as a matrix chopper (MC) or matrix-reactance chopper (MRC). A simplified schematic diagram of a hybrid transformer (HT) is shown in Fig. 6.6.

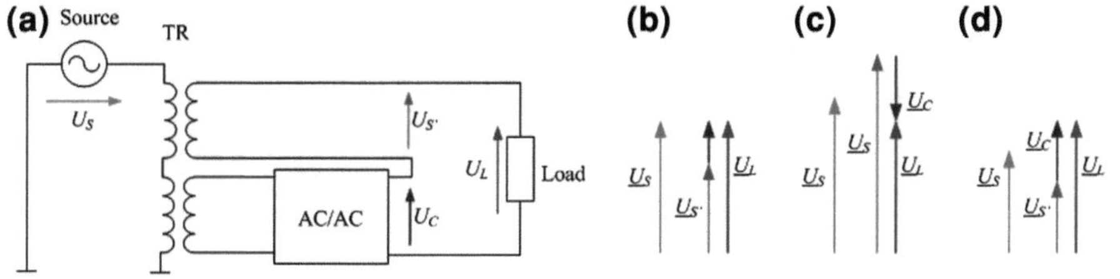

Fig. 6.6 Hybrid transformer: **a** simplified schematic diagram; **b** voltage phasors during nominal conditions, **c** during voltage swell and **d** during voltage sag

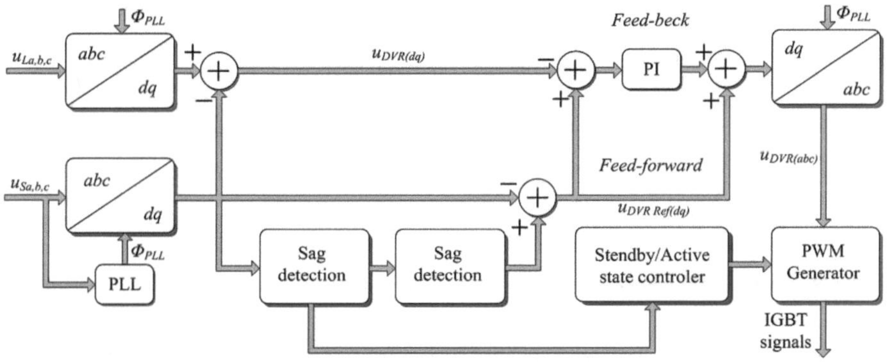

Fig. 6.7 Simplified schematic block diagram of a DVR control unit

6.2 Identification of Voltage Variation

A very important feature of voltage compensators is rapid dynamic reaction speed and quick response to source voltage changes. Several control techniques have been proposed for voltage sag compensation, such as the pre-sag method, the in-phase method and minimal energy control [16, 17]. In the pre-sag compensation technique, the DVR supplies the difference between the sagged and pre-sagged voltage and restores the voltage magnitude and the phase angle to the nominal pre-sag condition. In the case of the in-phase compensation method, only the voltage magnitude is compensated. Voltage U_C of DVR is in-phase with the source voltage U_S. This method minimizes the voltage injected by the DVR, unlike in the pre-sag compensation. Pre-sag compensation and in-phase compensation must inject active power to loads almost all the time. The fundamental idea of the energy optimization method is to make injection active power zero. In order to minimize the use of real power the voltages are injected at a 90° phase angle to the supply current [17]. An exemplary simplified schematic block diagram of the control strategy of a DVR device based on the positive and negative rotating ***dq***-reference frames is shown in Fig. 6.7 [10, 15, 18]. This is one of the many control techniques of a DVR.

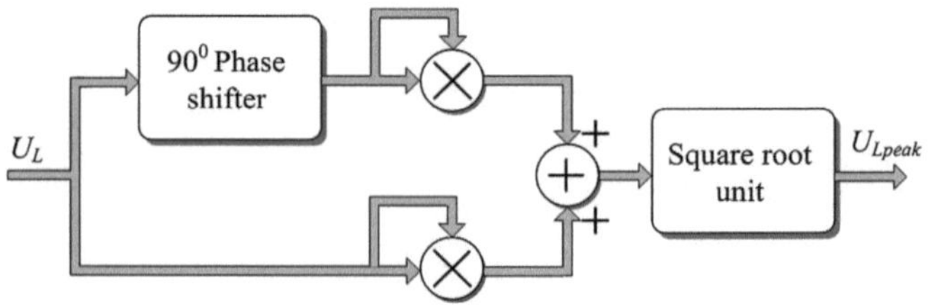

Fig. 6.8 Schematic block diagram of peak detector

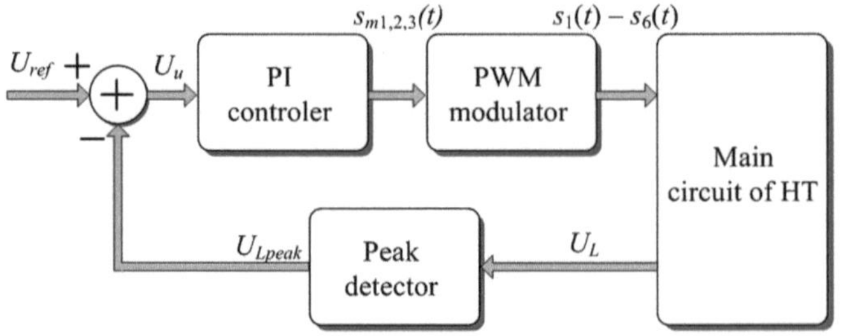

Fig. 6.9 Simplified schematic block diagram of the control circuit of HT

As can be seen in Fig. 6.7 the control technique is based on ***abc/dq*** transformation. The voltage sag detection method is based on the RMS of the error vector, which allows for detection of symmetrical and asymmetrical sags, as well as the associated phase jump. To generate IGBT signals a conventional PWM method, a hysteresis voltage controller method or a space vector method can be used.

In the case of HT using MC or MRC, identification of the voltage amplitude changes is realized by a peak detector [7]. The function realized by a peak detector is described in Eq. 6.5:

$$u_{L\mathrm{peak}} = \sqrt{(U_L \sin(\omega t))^2 (U_L \cos(\omega t))^2} \tag{6.5}$$

Schematic block diagram of a peak detector unit is shown in Fig. 6.8.

The input signal of the peak detector—load voltage U_L—is distributed to the two signals. One of them is shifted about $\pi/2$ according to (6.5). Both signals are squared and then are summed. Finally, to obtain the output signal $U_{L\mathrm{peak}}$, the square root of this sum is obtained.

An exemplary simplified schematic block diagram of the HT control unit is shown in Fig. 6.9 [28].

As shown in Fig. 6.9 the peak detector unit is located in the feedback loop coupling. The input signal to the peak detector unit is the load voltage U_L (6.5). In the main track of the control circuit are placed the controller *PI*, and the modulator PWM. A more detailed schematic diagram of the control unit of the HT is shown in Fig. 6.10 [9, 28].

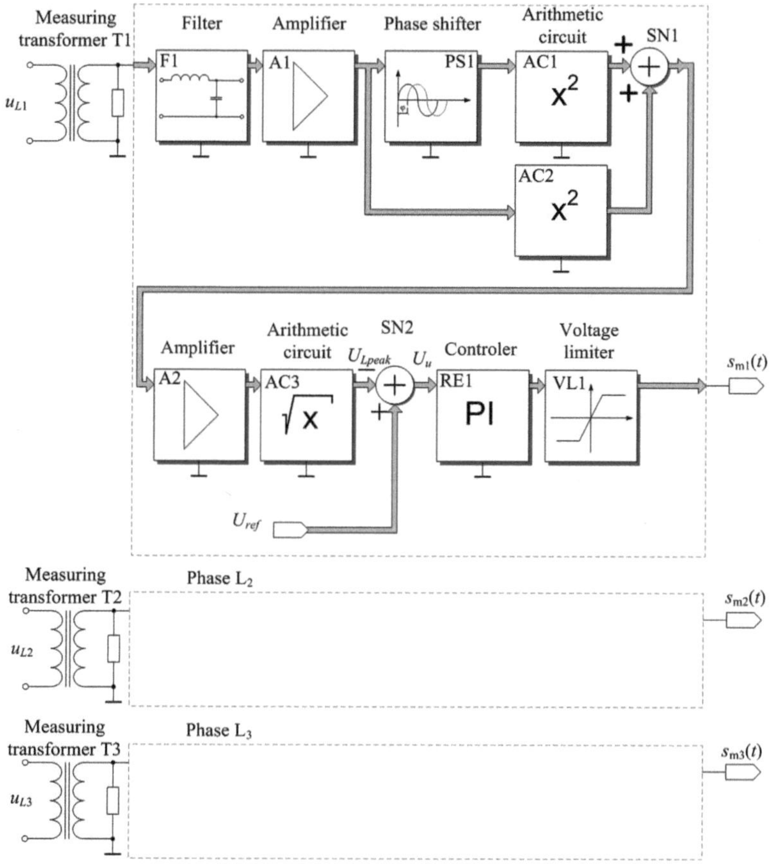

Fig. 6.10 Schematic block diagram of the control circuit of a three-phase SVC

The output of the peak detector unit is connected to the summation node when the reference signal U_{ref} (Figs. 6.9 and 6.10) has been subtracted from it. The error signal from the summation node is fed to the *PI* controller input. The amplifiers *A*1 and *A*2 realize the function of matching signal levels. The time response of the control circuit to source voltage changes is dependent on the *PI* controller parameters. The control circuit of the HT provides a fast response of the HT during voltage sags and swells and satisfactory dynamics of the presented HT.

6.3 Series Voltage Compensators

A DVR is one of the power electronic devices which are connected to the distribution system in series. A simplified schematic diagram of a three-phase DVR is shown in Fig. 6.11. The main advantage of a DVR is that under nominal voltage

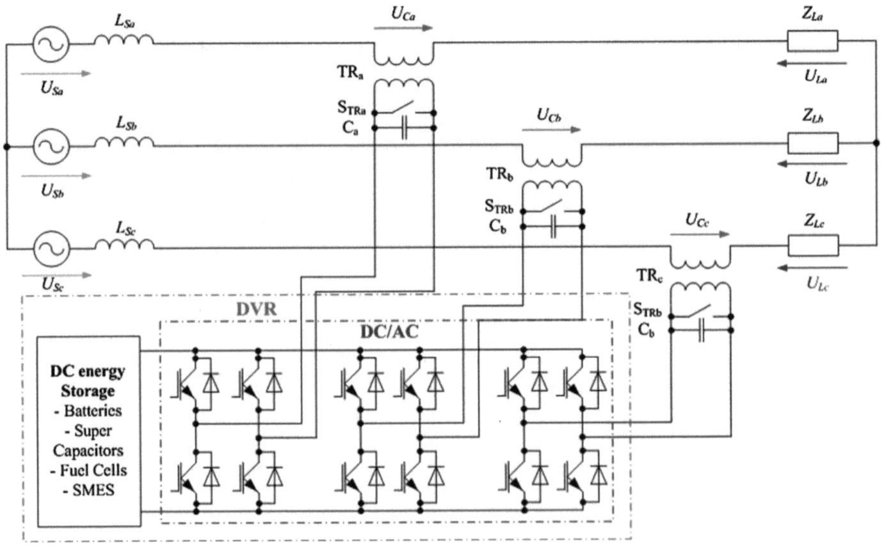

Fig. 6.11 Simplified schematic diagram of three-phase DVR

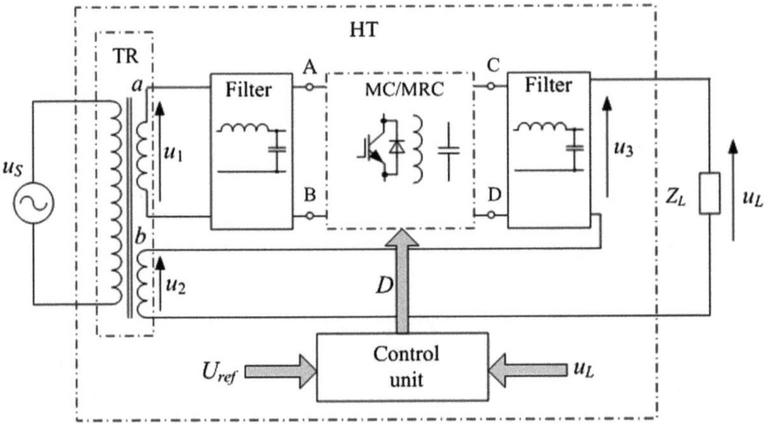

Fig. 6.12 Simplified schematic diagram of single-phase HT

conditions the DC/AC converter is not working. At this time the series transformers $TR_{a,b,c}$ are shorted by switches $S_{TRa,b,c}$ (Fig. 6.11). The DVR is activated during voltage sags or swells. This is a very important feature of a DVR. Another advantage is its capability to compensate deep voltage sags and swells, even short voltage interruptions.

As shown in Fig. 6.11 the DVR has no galvanic separation between source and load. The energy is transmitted from source (U_{Sa}, U_{Sb}, U_{Sc}) to the load (Z_{La}, Z_{Lb} and Z_{Lc}) directly. Energy from the DC energy storage to the load is transmitted during voltage sags or swells.

Galvanic separation between source and load is provided in the HT topology (Fig. 6.12).

As shown in Fig. 6.12 the presented solution has two main units: the first one is a conventional electromagnetic transformer TR with two secondary windings *a* and *b*; the second one is a PWM AC/AC (MC, MRC). The circuit shown in Fig. 6.12 has two couplings, an electromagnetic one realized by means of the TR and an electric one realized by means of the MC or MRC, therefore called a HT.

The main secondary winding *a* of TR is connected by input *LC* filter with AC/AC converter. Secondary winding *b* is series connected with output of AC/AC converter and load. The voltage ratios of the transformer secondary windings *a* and *b* are equal $n_a = 4/3$ and $n_b = 2/3$, respectively. Output voltage of HT U_L is an algebraic sum of the output voltage of AC/AC converter U_3 and secondary voltage of TR U_2. The topologies of the MC and MRC, which are realized by an electrical coupling in the HT circuit, are described in [23–26]. The MC and MRC have various properties, especially voltage gain (voltage transmittance). By using a different topology of MC and MRC in constructing a HT it is easy to obtain different properties from the HT. Schematic diagrams of selected MCs and MRCs with voltage transmittances (voltage gain) characteristics are presented in Table 6.1.

Idealized voltage transmittances of selected HTs are presented in Table 6.2.

The voltage transmittance of a HT in relation to the voltage ratio of TR is shown in Figs. 6.13, 6.14 and 6.15 [28].

As clear in Figs. 6.13–6.15 the range of change of voltage transmittance of a HT is dependent on the MC/MRC topology, pulse duty factor *D* and TR voltage ratio value.

A schematic diagram of a single-phase HT using MC is shown in Fig. 6.16.

The MC is supplied from one of two secondary windings of TR. The output voltage is the sum of the main voltage U_2 and the output voltage of the MC (adjustable voltage). Exemplary idealized voltage time waveforms and phasors illustrating the operation of the presented circuit are shown in Fig. 6.17.

The RMS value of the output voltage of the MC U_3 is dependent on the duty pulse factor described by Eq. 6.6.

$$D = \frac{t_{on}}{T_S} \tag{6.6}$$

Taking into account (6.6) the voltage transmittance of the MC in Fig. 6.16 can thus be written as :

$$\underline{H}_U^{MC} \approx \frac{\underline{U}_3}{\underline{U}_1} \approx \frac{\underline{U}_1 \cdot D}{\underline{U}_1} \approx D \tag{6.7}$$

Taking into account the voltage ratios of the TR (n_a, n_b), and dependences (6.6) and (6.7) the voltage transmittances of the HT in Fig. 6.17 is described as:

$$\underline{H}_U^{TH_{SM}} \approx \frac{\underline{U}_L}{\underline{U}_S} \approx \frac{n_b \underline{U}_S + D n_a \underline{U}_S}{\underline{U}_S} \approx n_b + D n_a \approx \frac{2}{3} + D\frac{4}{3} \tag{6.8}$$

Table 6.1 Schematic diagrams of MCs and MRCs used to construct HT

	Unipolar MC (MC-u)	Bipolar MC (MC-b)

1

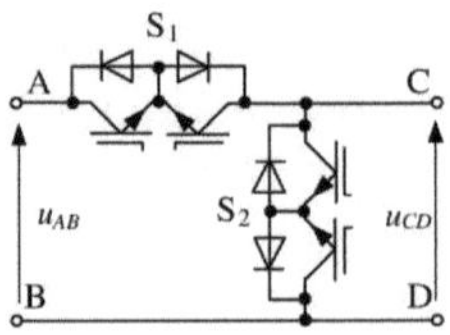

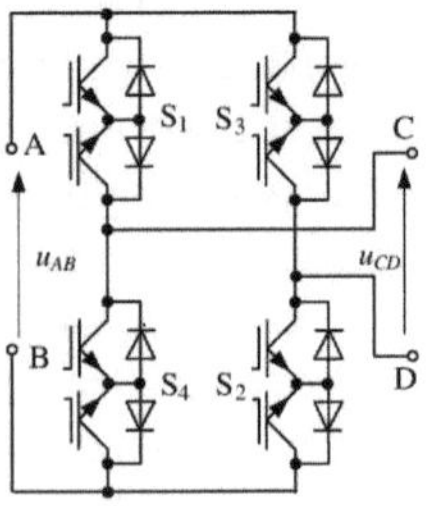

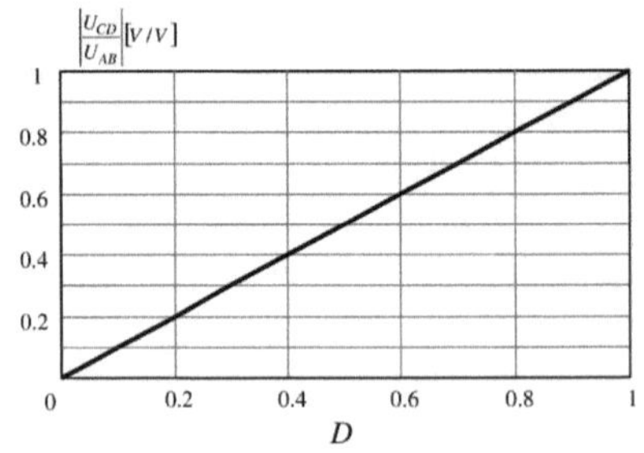

Boost

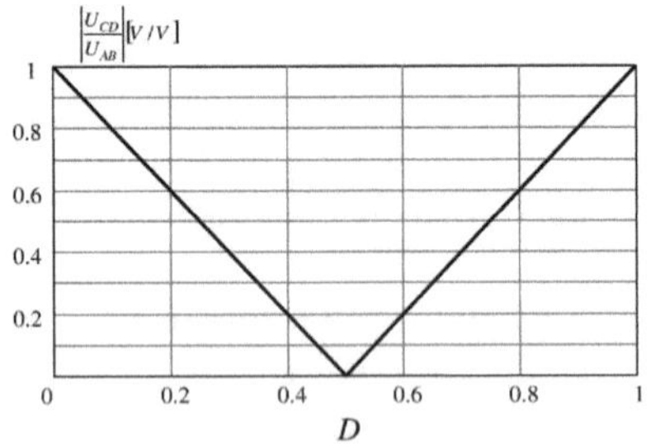

Buck – Boost

2

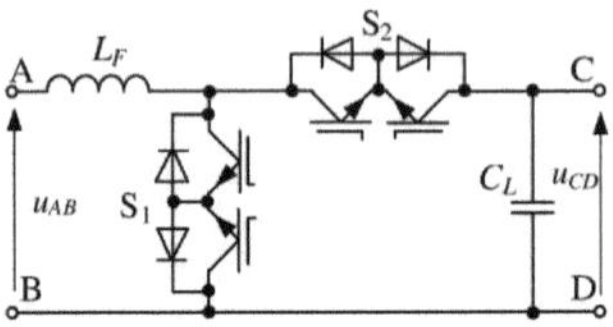

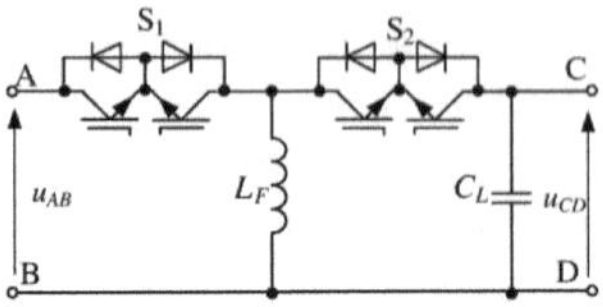

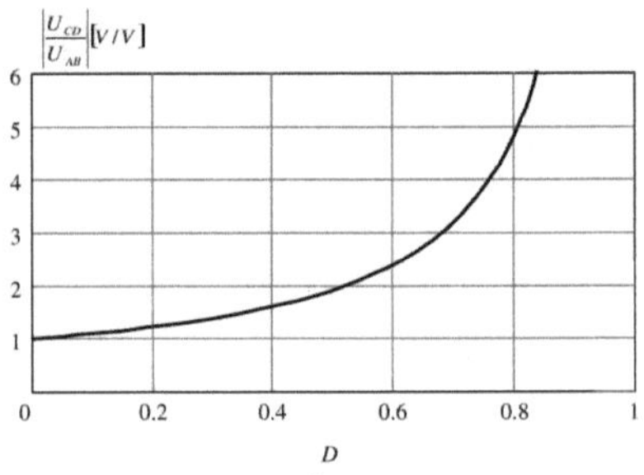

Ćuk

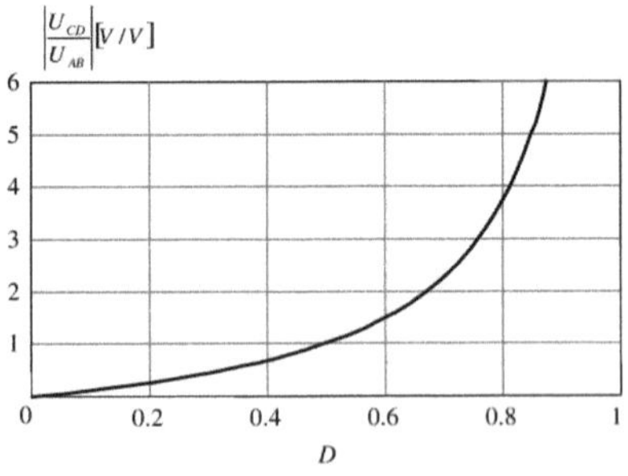

Ćuk B2

(continued)

Table 6.1 (continued)

<table>
<tr><th></th><th>Unipolar MC (MC-u)</th><th>Bipolar MC (MC-b)</th></tr>
<tr><td>3</td><td>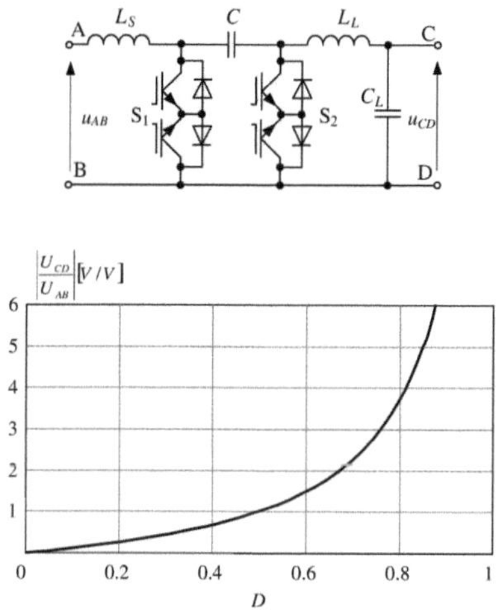
</td><td>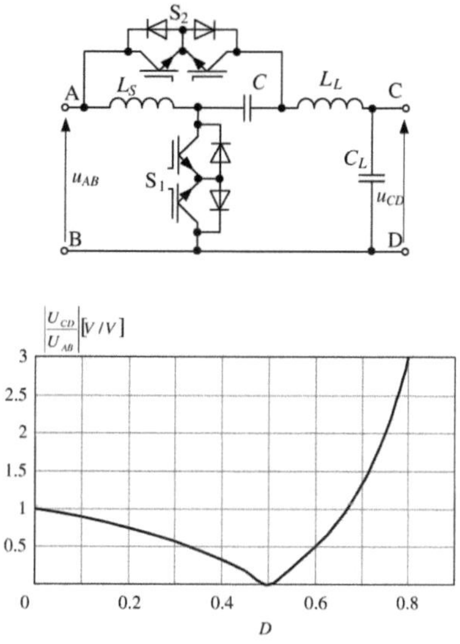
</td></tr>
<tr><td></td><td>Zeta</td><td>SEPIC</td></tr>
<tr><td>4</td><td>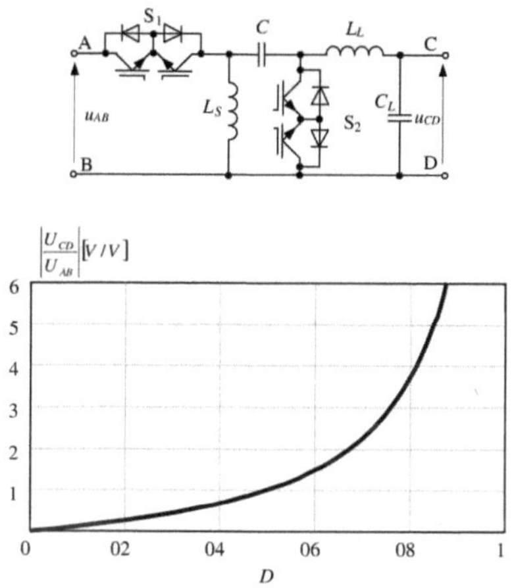
</td><td>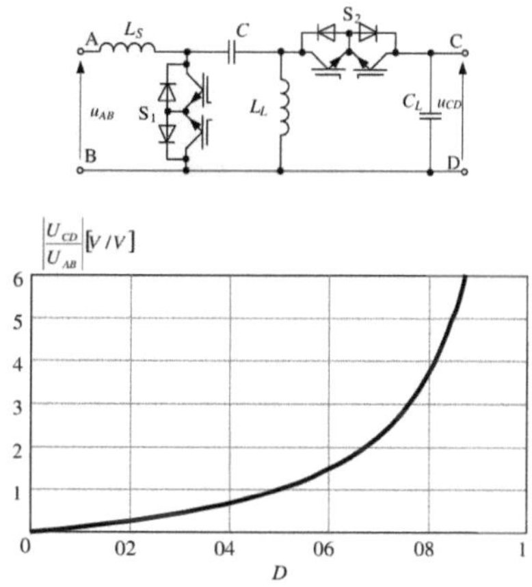
</td></tr>
</table>

The idealized static characteristics of voltage transmittance as a function of duty pulse factor D of the HT in Fig. 6.16 is shown in Fig. 6.18.

As evident in Fig. 6.19, the output voltage of the HT is less than the source voltage for $0 \leq D \leq 0.25$. For a value of D in the range of 0.25–1 the output voltage of the HT is greater than the supply voltage.

Every single-phase circuit can be developed into a three-phase topology. A simplified schematic diagram of a three-phase HT is shown in Fig. 6.19.

Table 6.2 Idealized voltage transmittances of selected HT

HT using	H_U[V/V]
MC-u	$n_a D + n_b$
MC-b	$n_a(2D - 1) + n_b$
MRC Boost	$\frac{n_a D}{1 - D} + n_b$
MRC Buck-Boost	$\frac{n_a D}{1 - D} + n_b$
MRC Ćuk	$\frac{n_a D}{1 - D} + n_b$
MRC Ćuk B2	$\frac{n_a(1 - 2D)}{(1 - D)} + n_b$
MRC Zeta	$\frac{n_a D}{1 - D} + n_b$
MRC SEPIC	$\frac{n_a D}{1 - D} + n_b$

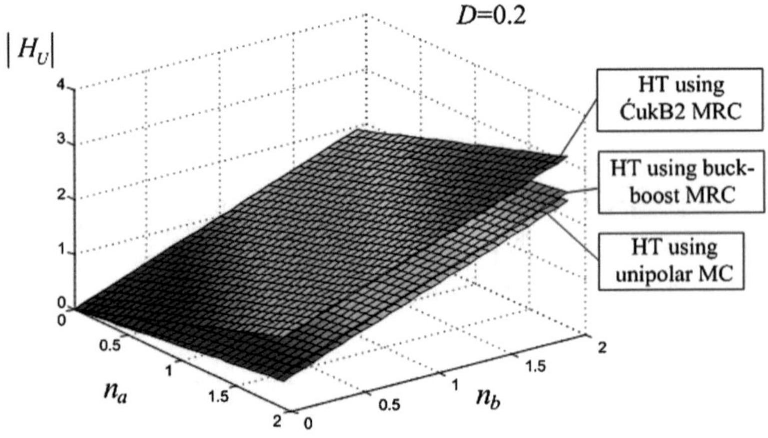

Fig. 6.13 The voltage transmittance of HT in relation to voltage ratio of TR for $D = 0.2$

Similar to the single-phase TH topologies, the three-phase HT contains two main units. The first one is a three-phase transformer effecting an electromagnetic coupling and the second one is a MC or MRC effecting an electrical coupling (Fig. 6.19). Construction of the three-phase transformer is different in comparison with a conventional three-phase transformer. The transformer TR has two secondary windings in each phase (a_{1-2}, a_{3-4}, b_{1-2}, b_{3-4}, c_{1-2}, c_{3-4}) (Fig. 6.20).

With regard to the operating conditions of the transformer it is possible to work with various configurations of windings in transformer TR (Yy0, Yy6, Dy11, Dy5 Dz6, Dz0, Yz11, Yz5). The windings configuration has no influence on the HT properties. The properties of the HT are dependent on the MC or MRC topology connected to the HT circuit and the voltage ratio of the TR (as in the case of

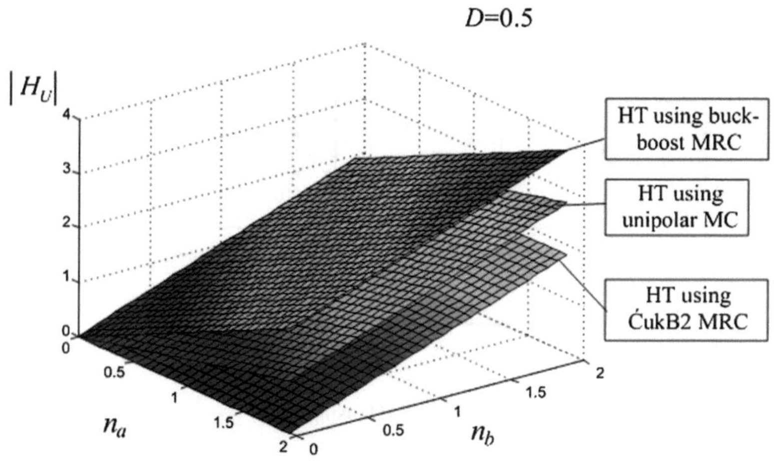

Fig. 6.14 The voltage transmittance of HT in relation to voltage ratio of TR for $D = 0.5$

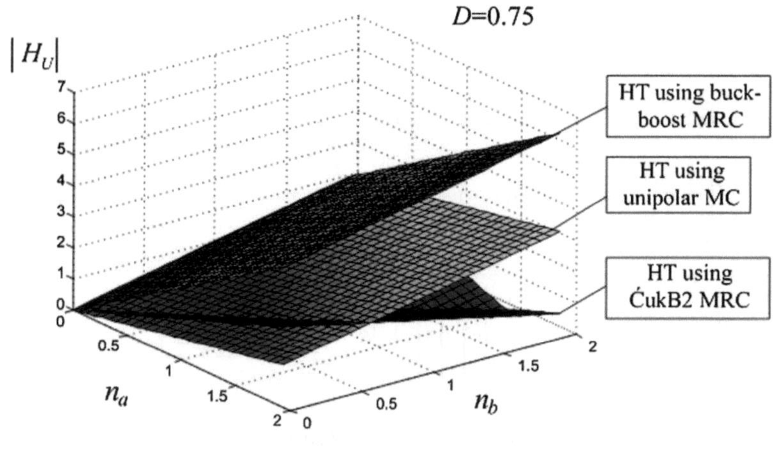

Fig. 6.15 The voltage transmittance of HT in relation to voltage ratio of TR for $D = 0.75$

single-phase HT). The HT could operate both in a three-phase three-wire system and in a three-phase four-wire system [28].

A schematic diagram of a three-phase HT using buck-boost MRC (MRC *b–b*) is shown in Fig. 6.21.

As shown in Fig. 6.21 the primary windings are in Y-configuration. The main secondary windings (a_1, a_2, a_3) of the TR also have a *Y* configuration and, by input filter *LC*, are connected to MRC *b–b*. The secondary phase windings (b_1, b_2, b_3) are connected in series with the required phase output connectors of the MRC. The output voltages of the HT (u_{L1}, u_{L2}, u_{L3}) are the sum of the secondary voltages ($u_{S1}n_b$, $u_{S2}n_b$, $u_{S3}n_b$) and the phase output voltages of the MRC. The voltages of the

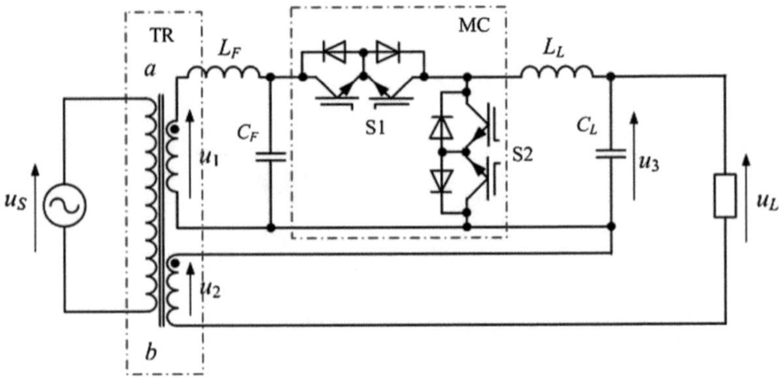

Fig. 6.16 Schematic diagram of single-phase HT using MC

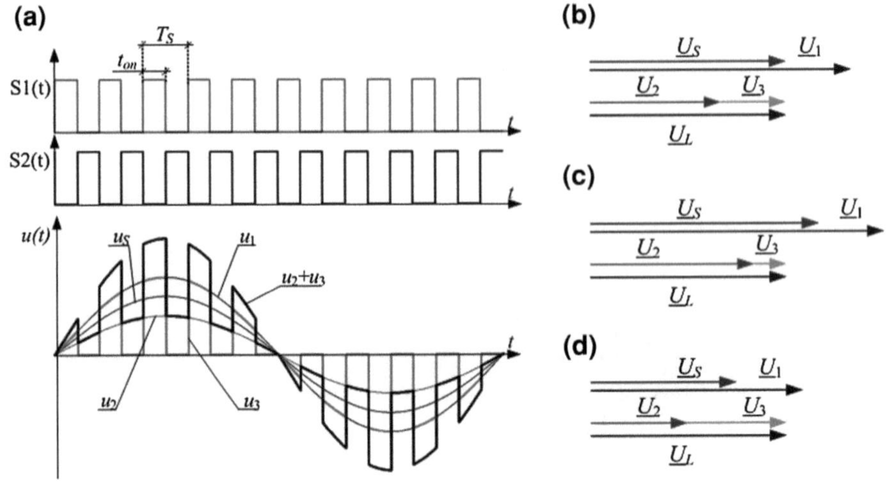

Fig. 6.17 Exemplary idealized voltages of HT (Fig. 6.16): **a** time waveforms; **b** phasors during nominal conditions, **c** during voltage swell, **d** during voltage sag

transformer secondary windings a_1, a_2, a_3 and b_1, b_2, b_3 equal $n_a = 4/3$ and $n_b = 2/3$, respectively [9, 27, 28]. Exemplary voltage phasors of the presented HT are shown in Fig. 6.22.

$$\begin{aligned} \underline{U}_{S1} &= U_S \cdot e^{j0} \\ \underline{U}_{S2} &= U_S \cdot e^{-j\frac{2\pi}{3}} \\ \underline{U}_{S3} &= U_S \cdot e^{j\frac{2\pi}{3}} \end{aligned} \tag{6.9}$$

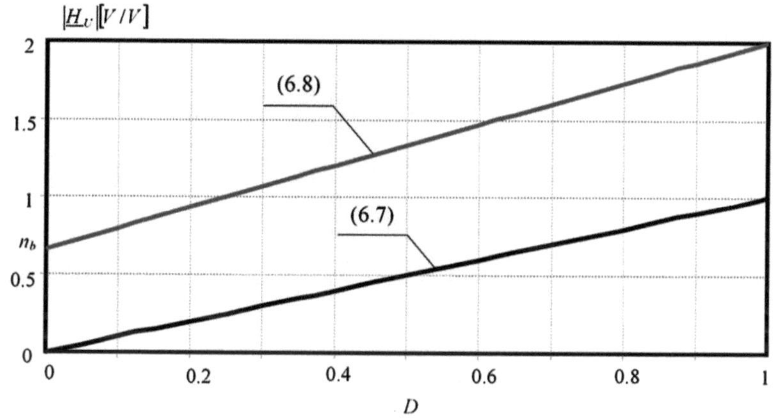

Fig. 6.18 Idealized static characteristics of voltage transmittance as a function of duty pulse factor D

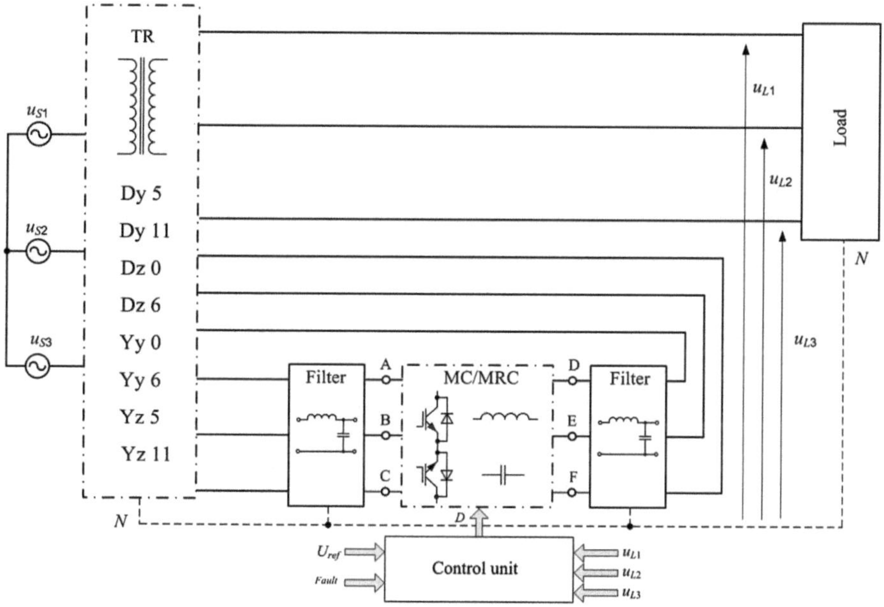

Fig. 6.19 Simplified schematic block diagram of three-phase HT

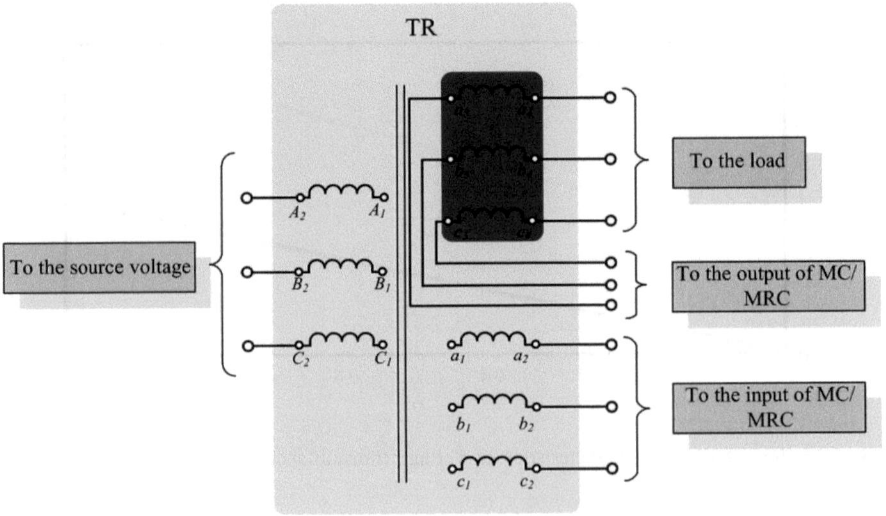

Fig. 6.20 Schematic diagram of transformer TR in three-phase HT circuit

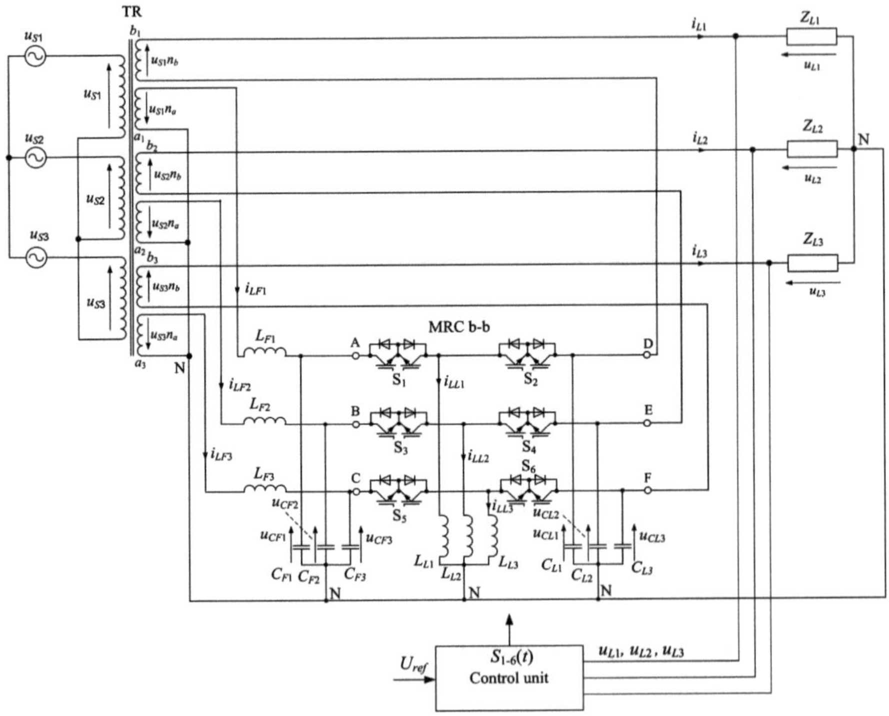

Fig. 6.21 Schematic diagram of a three-phase HT using buck-boost MRC

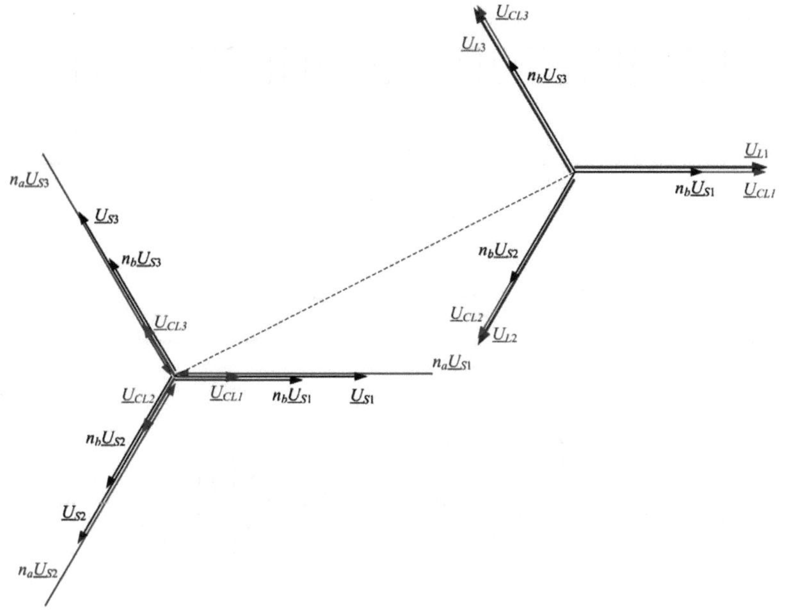

Fig. 6.22 Exemplary voltage phasors of HT using buck-boost MRC

$$\begin{aligned} n_a\underline{U}_{S1} &= -n_a U_S \cdot e^{j0} \\ n_a\underline{U}_{S2} &= -n_a U_S \cdot e^{-j\frac{2\pi}{3}} \\ n_a\underline{U}_{S3} &= -n_a U_S \cdot e^{j\frac{2\pi}{3}} \end{aligned} \tag{6.10}$$

$$\begin{aligned} n_b\underline{U}_{S1} &= n_b U_S \cdot e^{j0} \\ n_b\underline{U}_{S2} &= n_b U_S \cdot e^{-j\frac{2\pi}{3}} \\ n_b\underline{U}_{S3} &= n_b U_S \cdot e^{j\frac{2\pi}{3}} \end{aligned} \tag{6.11}$$

$$\begin{aligned} \underline{U}_{CL1} &= n_a U_S H_U^{b-b} \cdot e^{j0} \\ \underline{U}_{CL2} &= n_a U_S H_U^{b-b} \cdot e^{-j\frac{2\pi}{3}} \\ \underline{U}_{CL3} &= n_a U_S H_U^{b-b} \cdot e^{j\frac{2\pi}{3}} \end{aligned} \tag{6.12}$$

$$\begin{aligned} \underline{U}_{L1} &= n_a\underline{U}_{S1} + \underline{U}_{CL1} \\ \underline{U}_{L2} &= n_a\underline{U}_{S2} + \underline{U}_{CL2} \\ \underline{U}_{L3} &= n_a\underline{U}_{S3} + \underline{U}_{CL3} \end{aligned} \tag{6.13}$$

where $U_{S1,2,3}$—phase voltages in complex form, U_S—maximum value of source voltage, n_a, n_b—voltage ratio a and b of transformer TR, $U_{L1,2,3}$—load voltages in

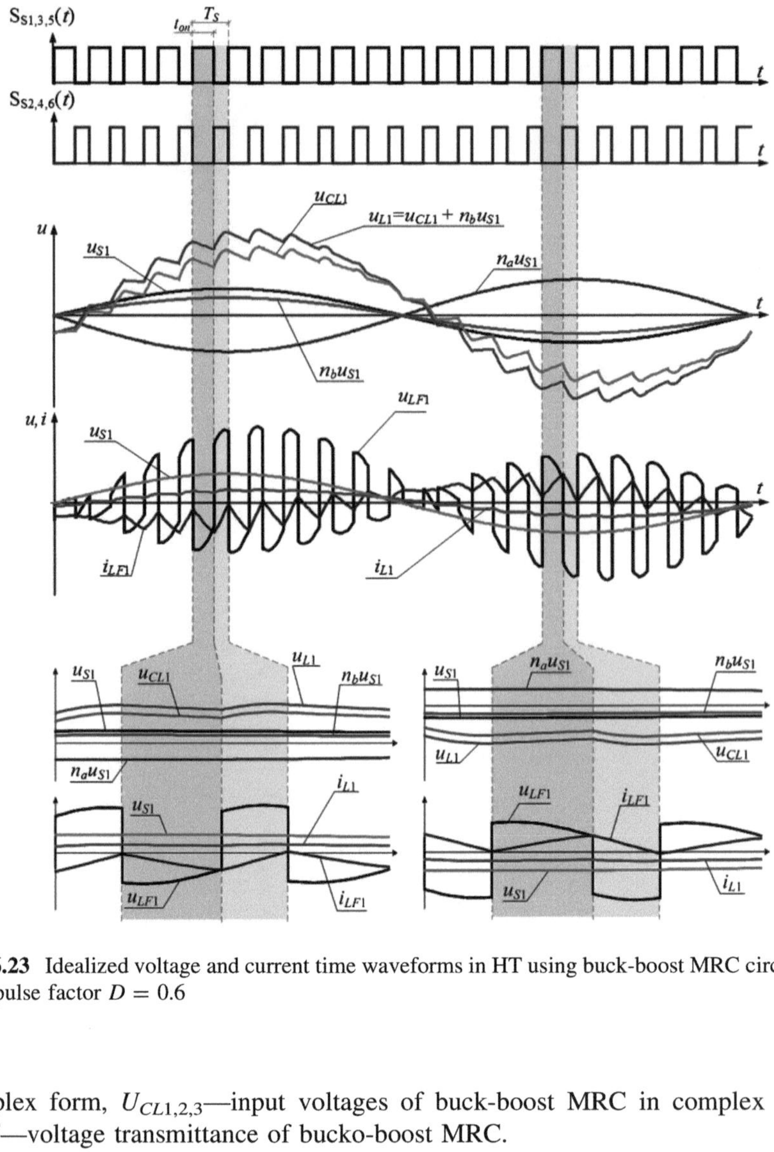

Fig. 6.23 Idealized voltage and current time waveforms in HT using buck-boost MRC circuit for duty pulse factor $D = 0.6$

complex form, $U_{CL1,2,3}$—input voltages of buck-boost MRC in complex form, H_U^{b-b}—voltage transmittance of bucko-boost MRC.

$$H_U^{b-b} = \frac{\underline{U}_{CL}}{\underline{U}_{CF}} \approx \frac{D}{1-D} \tag{6.14}$$

Idealized time voltage waveforms of the HT with MRC *b–b* for two different values of pulse duty factor *D* are shown in Figs. 6.23 and 6.24.

Experimental time waveforms of voltage load in a three-phase transformer (Fig. 6.22) during source voltage sag and swell are shown in Figs 6.25, 6.26 and 6.27.

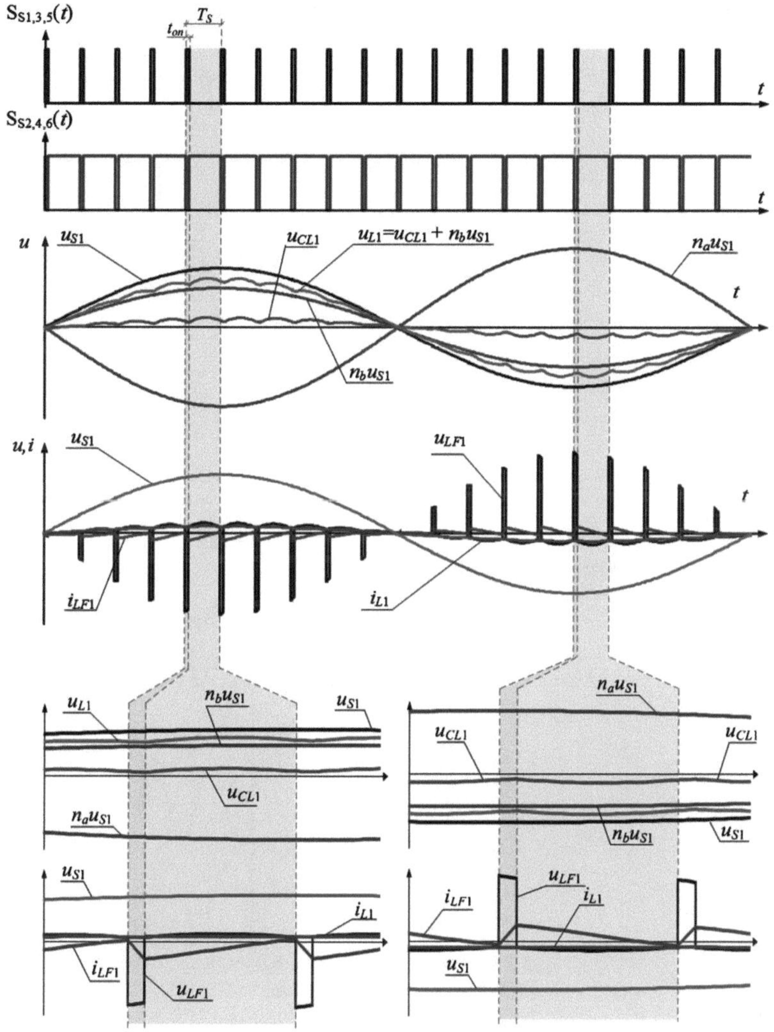

Fig. 6.24 Idealized voltage and current time waveforms in HT using buck-boost MRC circuit for duty pulse factor $D = 0.1$

As can be seen in Figs. 6.25–6.27 the output voltages $U_{L1,2,3}$ have a constant value throughout the voltage sag and swell duration. The transient state during source voltage step up/dawn is less than 10 ms. The static character of the magnitude of voltage transmittance as a function of pulse duty factor D is shown in Fig. 6.28.

As is clear in Fig. 6.28 the output voltage of the considered HT is less, approximately equal to or greater, than the source voltage for $D < 0.2$, for

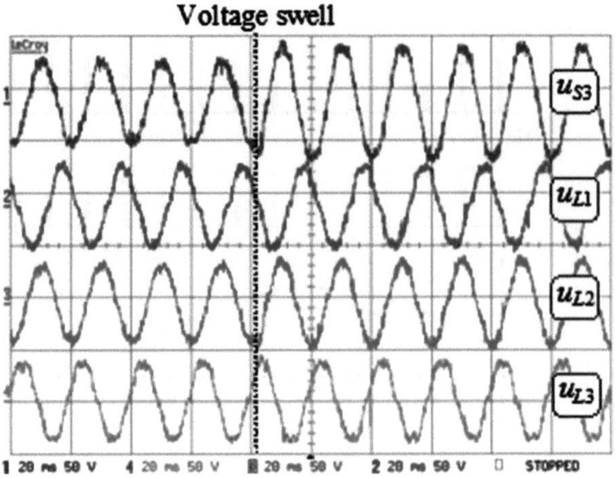

Fig. 6.25 Experimental source and load voltage time waveforms during 140% U_S voltage swell

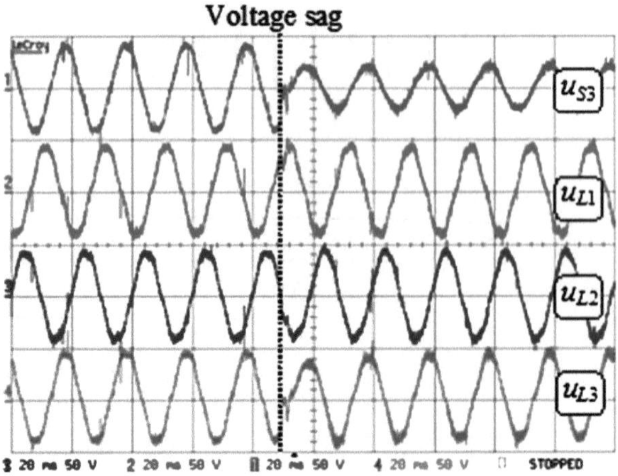

Fig. 6.26 Experimental source and load line voltage time waveforms during 45% U_S voltage sag

$D = 0.2$ and for $D > 0.2$, respectively. The range of change of output voltage is from $0.66U_S$ to more than $3U_S$.

For this reason, the MRC is working all the time. During nominal conditions about 30% of all energy is transmitted to the load by the MRC and the rest is transmitted by the second winding of the TR (Fig. 6.29).

The buck-boost MRC allows the possibility of bidirectional transfer of energy (from source to load and from load to the source). These properties of the MRC, when using a special control strategy, allow power flow control in the power system.

Fig. 6.27 Experimental source and load line voltage time waveforms during U_S voltage sag

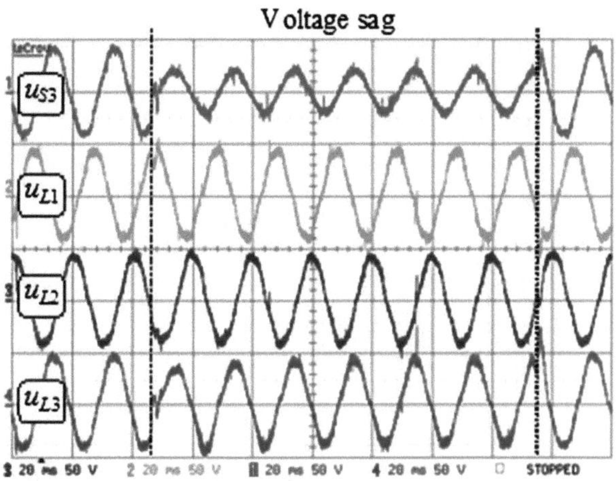

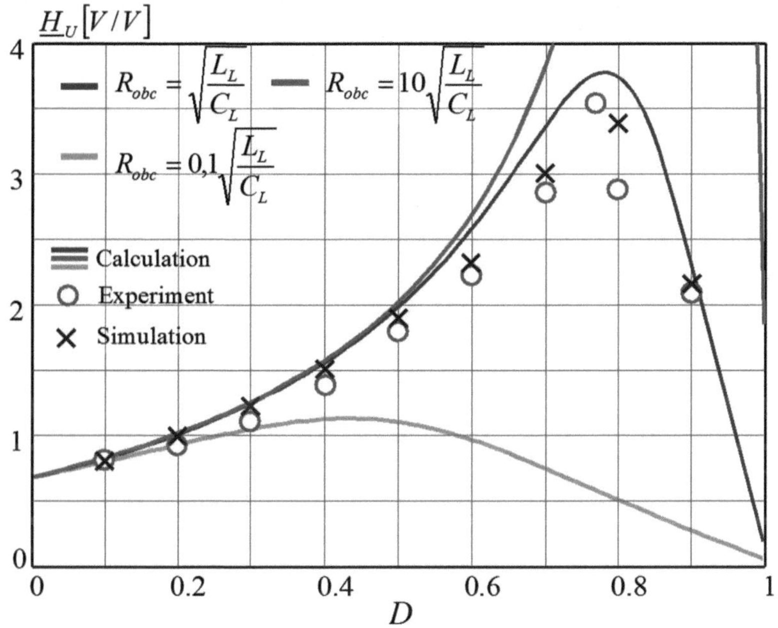

Fig. 6.28 Static character of magnitude of voltage transmittance HT as a function of the pulse duty factor D

To use the DC energy storage in the HT the DC link in the AC/AC converter is necessary. The conception of the HT with a DC energy storage is shown in Fig. 6.30. The AC/DC/AC converter is supplied from one of two three-phase

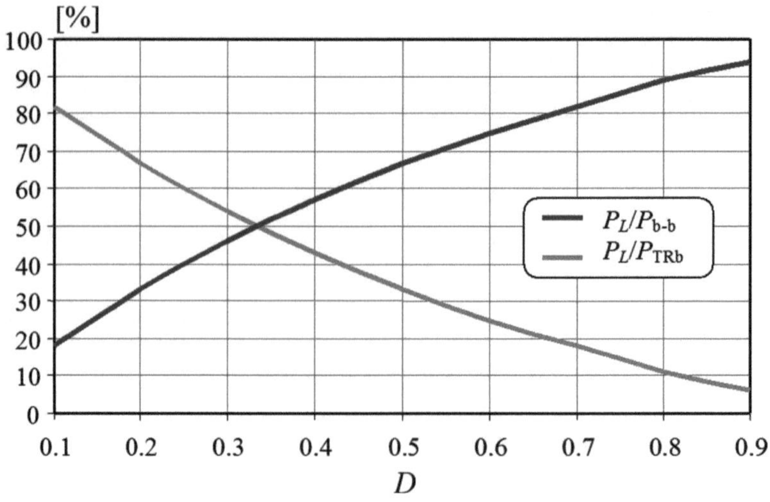

Fig. 6.29 Estimated relations between energy transmitted to the load by MRC and TR in comparison to total load energy

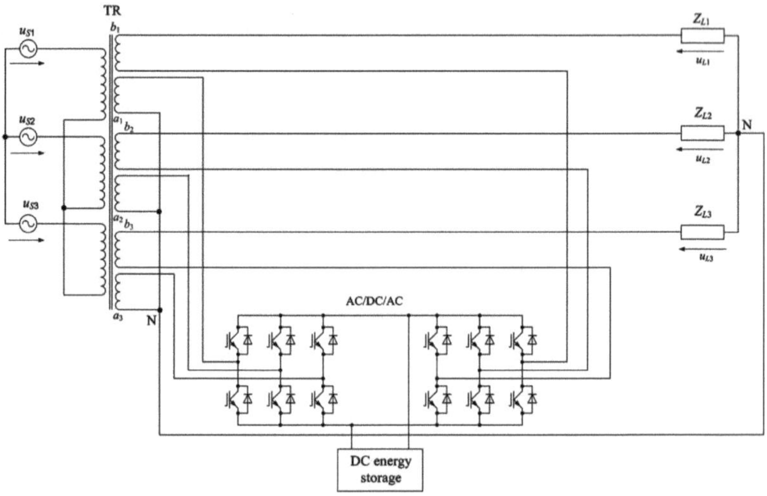

Fig. 6.30 Simplified schematic diagram of three-phase HT using AC/DC/AC converter with energy storage unit

secondary windings of transformer TR (a_1, a_2, a_3). The output of the AC/DC/AC converter is connected in series with the main three-phase secondary windings of TR (b_1, b_2, b_3).

6.4 Summary

Series voltage compensators are power electronics devices to mitigate unwanted effects on the consumer side, such as voltage sags, swells and interrupts. The most common device among them are DVR. The main advantage of these devices is the ability to compensate deep voltage sags and short interrupts. Moreover, during nominal conditions the DVR is in an idle state. The DVR is actuated by the detection of a voltage change (sag, overvoltage, interrupt). This is a very important feature from the point of view of the input power factor and efficiency. The disadvantage of the DVR is the necessity to use a DC energy storage device (batteries, super capacitors, fuel cells, etc.).

The HT is a kind of special SVC. In comparison with a conventional solution with series transformer (e.g. DVR), the HT, by using a conventional transformer TR, provides galvanic separation between source voltage and load. Moreover the HT allows compensation of long voltage sags and swells. In the case of the DVR the operating time is limited and depends on the duration and value of the voltage change and power load. In the case of the HT, the operating time is independent of voltage sag/swell duration. Another advantage of the HT is galvanic separation between source and load and the redundancy of any DC energy storage. The main disadvantage of the HT is the inability to compensate voltage interruption.

Series voltage compensators (DVR, HT) are very important parts of a power system and allow the control (stabilization) of voltage values in the power system grid. Moreover, by using a special control strategy they allow the possibility to control the power flow in a grid.

References

1. Standard EN 50160, Voltage characteristics of public distribution systems
2. Conrad L, Little K, Grigg C (1991) Predicting and preventing problems associated with remote fault—clearing voltage dips. IEEE Trans Ind Appl 27(1):167–172
3. Milanowić J, Hiskansen I (1995) Effect of load dynamics on power system damping. IEEE Trans Power Syst 10(2):1022–1028
4. Djokic Z, Desment J, Vanalme G, Milanovic J, Stockman K (2005) Sensitivity of personal computer to voltage sags and short interruptions. IEEE Trans Power Delivery 20(1):375–383
5. Duran-Gomez J, Prased P, Enjeti N, Woo B (1999) Effect of voltage sags on adjustable-speed drives: a critical evaluation on an approch to improve performance. IEEE Trans Ind Appl 35(6):1440–1449
6. Falce A, Matas G, Da Silva Y (2004) Voltage sag analysis and solution for an industrial plant with embedded induction motors. Ind Appl Conf 4:2573–2578
7. Chu HY, Jou HL, Huang L (1992) Transient response of a peak voltage detector for sinusoidal signals. IEEE Trans Ind Electron 39(1):74–79
8. Fedyczak Z, Kaniewski J, Klytta M (2007) Single-phase hybrid transformer using matrix-reactance hopper with ćuk topology. In: EPE 07 conference, Denmark, Aalborg
9. Kaniewski J, Fedyczak Z (2011) Modeling and analysis of dynamic properties of the hybrid transformer with MRC. Electr Rev 2011(1):45–50

10. Delfino B, Fornari F, Procopio R (2005) An effective SSC control scheme for voltage Sag Compensation. IEEE Trans Power Delivery 20(3):2100–2107
11. Choi SS, Li JD, Mahinda Vilathgamuwa D (2005) A Generalized Voltage compensation Strategy for Mitigating the Impacts of Voltage Sags/Swells. IEEE Trans Power Delivery 20(3):2289–2297
12. Woodley NH, Morgan L, Sundaram A (1999) Experience with an inverter—based dynamic voltage restorer. IEEE Trans Power Delivery 14(3):1181–1186
13. Benysek G (2007) Improvement in the quality of Delivery of Electrical Energy using PowerElectronics Systems. Springer-Verlag, London
14. Hingorani NG, Gyugyi L (1999) Understanding FACTS: concepts and technology of flexible ac transmission systems. IEEE Press, New York
15. Nielsen JG, Newman M, Nielsen H, Blaabjerg F (2004) Control and testing of a dynamic voltage restorer (DVR) at medium voltage level. IEEE Trans Power Electron 19(3):806–813
16. Meyer C, Romaus C, Doncker RW (2005) Optimized controlstrategy for a medium-voltage DVR. In: Powerelectronics specialists conference, IEEE, pp 1887–1993
17. Kim H (2002) Minimal energy control for a dynamic voltage restorer. In: PCC conference, IEEE, vol 2, pp 428–433
18. Ezoji H, Sheikholeslami A, Tabasi M, Saeednia MM (2009) Simulation for dynamic voltage restorer Using Hysteresis VoltageControl. Eur J Sci Res 27(1):152–166
19. Aeloiza E, Enjeti P, Moran L, Pite I (2003) Next generation distribution transformer: to address power quality for critical loads. In: PESC'03 IEEE conference, vol 3, pp 1266–1271
20. Kaniewski J, Fedyczak Z (2009) Modelling and analysis of three-phase hybrid transformer using matrix-reactance chopper. Electr Rev 2:100–105
21. Kaniewski J, Fedyczak Z, Klytta M, Łukiewski M, Szcześniak P (2009) Implementation of a three-phase hybrid transformer using a matrix chopper. In: Proceedings of the EPE 09 conference
22. Fedyczak Z, Kaniewski J (2007) Modeling and analysis of three-phase hybrid transformer using matrix converter. In: Proceedings of the CPE 2007 conference
23. Fedyczak Z (2006) PWM AC voltage transforming circuits (in Polish). University of Zielona Góra Press, Zielona Góra
24. Fedyczak Z (2003) The steady state basic energetic properties of PWMPWM AC line bipolar matrix-reactance choppers. Arch Electr Eng 52(3):303–316
25. Fedyczak Z (2001) Four-terminal chain parameters of averaged AC models of non-isolated matrix-reactance PWM AC line conditioners. Arch Electr Eng L(4):395–409
26. Fedyczak Z (2003) Steady state modelling of the bipolar PWM AC line matrix-reactance choppers based on ćuk topologies. Arch Electr Eng 2(3):303–316
27. Kaniewski J (2011) Modelling and analysis of three-phase hybrid transformer using buck-boost MR. In: Proceedings of the CPE 2011 conference
28. Kaniewski J (2011) Analysis and study the properties of hybrid transformer. Ph.D. thesis (in Polish), University of Zielona Góra Press, Zielona Góra

Index

G. Benysek and M. Pasko (eds.), *Power Theories for Improved Power Quality*, Power Systems, DOI: 10.1007/978-1-4471-2786-4,

MIX
Papier aus verantwortungsvollen Quellen
Paper from responsible sources
FSC® C105338
FSC
www.fsc.org

If you have any concerns about our products,
you can contact us on
ProductSafety@springernature.com

In case Publisher is established outside the EU,
the EU authorized representative is:
Springer Nature Customer Service Center GmbH
Europaplatz 3, 69115 Heidelberg, Germany

Printed by Libri Plureos GmbH
in Hamburg, Germany